Virtual and Physical Modeling for Engineering Design:

A Student Activity Manual

Virtual and Physical Modeling for Engineering Design:

A Student Activity Manual

Aaron C. Clark and Jeremy V. Ernst
North Carolina State University

Original Authors From VisTE:
Eric N. Wiebe, Aaron C. Clark, Jeremy V. Ernst, Miriam G. Ferzli,
Carl N. Blue, and Julie H. Petlick

Australia • Brazil • Japan • Korea • Mexico • Singapore • Spain • United Kingdom • United States

CENGAGE
Learning™

Virtual and Physical Modeling for Engineering Design: A Student Activity Manual

Aaron C. Clark and Jeremy V. Ernst

Vice President, Career and Professional Editorial: Dave Garza

Director of Learning Solutions: Sandy Clark

Senior Acquisitions Editor: James DeVoe

Managing Editor: Larry Main

Product Manager: Mary Clyne

Editorial Assistant: Cris Savino

Vice President, Career and Professional Marketing: Jennifer McAvey

Executive Marketing Manager: Deborah S. Yarnell

Marketing Manager: Jimmy Stephens

Associate Marketing Manager: Mark Pierro

Production Director: Wendy Troeger

Production Manager: Mark Bernard

Content Project Management: Pre-PressPMG

Art Director: Bethany Casey

Photo Researcher: Pre-PressPMG

Compositor: Pre-PressPMG

For product information and technology assistance, contact us at
Cengage Learning Customer & Sales Support, 1-800-354-9706

For permission to use material from this text or product, submit all requests online at **cengage.com/permissions**
Further permissions questions can be emailed to
permissionrequest@cengage.com

Library of Congress Control Number: 2009922560

ISBN-13: 978-1-435-43905-4

ISBN-10: 1-435-43905-8

Delmar
5 Maxwell Drive
Clifton Park, NY 12065-2919
USA

Cengage Learning products are represented in Canada by Nelson Education, Ltd.

For your lifelong learning solutions, visit **delmar.cengage.com**

Visit our corporate website at **www.cengage.com**

Printed in Canada
1 2 3 4 5 6 12 11 10 09

Contents

UNIT 4   NANOTECHNOLOGY

UNIT 5   BIOMETRICS

Virtual and Physical Modeling for Engineering Design:

A Student Activity Manual

UNIT 1
Communications Technology: Introduction to Visualization

Unit Overview

I. Introduction

Every day graphics are being used to guide decision making. These decisions may concern interpreting the results of basic research studies in high-energy physics, developing an understanding of trends in global warming, the structural safety of a new bridge design, or identifying potentially hostile aircraft during war. In all of these cases, graphics are being used to communicate information to one or more people who are attempting to solve problems. These problems may concern the immediate safety of people or they may affect the quality of people's lives 100 years from now.

This unit introduces how graphics can be used to communicate technical and scientific information. In doing so, this unit shows how the design process can be applied to creation of graphics used to communicate data-driven information (e.g., charts and graphs) or conceptual information about technological and scientific systems. This unit focuses on the design and creation of two-dimensional (2D) graphics, and Unit 2 will focus on three-dimensional (3D) graphics.

Why not just use words or numbers to communicate this information? Words and numbers are powerful ways of communicating. However, they are not the optimal method of communicating all types of information. Especially in cases where individuals are being asked to synthesize large amounts of data and to understand trends or identify unusual conditions, graphics can be a powerful tool. As powerful as graphics can be, they are a language and have to be crafted in the same way that words needs to be constructed into sentences and paragraphs. This unit will introduce you to how problem-solving heuristics can be applied to the design of graphics to communicate technical and scientific information. Later activities will apply these tools to the communication of different areas of technology. Let's start this unit by looking at an example from business and manufacturing.

You work for a biotechnology firm manufacturing pharmaceutical compounds used in a revolutionary new class of cancer-fighting drugs. Demand is so high that your company needs to expand its operations. Building a new facility offers the possibility of greatly expanding your production capacity. However, it could also sink your young company if you end up not being able turn a profit on your new facility. The bottom line: a lot rests on making the right decisions concerning this new manufacturing facility. Part of the decision-making process is determining where to locate the plant. Factors such as the quality of the labor force, transportation of raw materials and finished product, and climatic conditions all come into play in making this decision.

What role do climatic conditions play? Weather impacts the efficiency and effectiveness of transportation systems. How often is the weather extreme enough that workers can't drive their cars to work or shipments of raw material can't make it to the plant? Maybe a bigger long-term factor is going to be how much it is going to cost to heat or cool your facility. If you are engaged in bioprocess manufacturing, temperature control of your facility is critical not only for the comfort of your workers, but for management of your manufacturing processes. These heating and cooling costs may be a key factor in determining the overall cost of operations of your new facility. In this unit, you will look at how graphic visualization technology might be used to help make these crucial decisions. You will also be introduced to a problem-solving method that can be used to help decide on optimal graphic designs.

II. Unit Learning Goals

- To help you develop an understanding of the historical perspective of graphics as a communications technology.
- To learn the design process for graphic communication of technical and scientific information.
- To develop an awareness of both inadvertent and purposeful misrepresentation of information with graphics.

III. Projects

Introductory Projects

PROJECT 1: Interpreting Graphics

This project will provide you with the opportunity to explore how graphics are currently being used to convey technical and scientific information. It also allows you to rehearse the analytic process of decomposing graphics into their component parts and identifying the role of each component. This analytic process also allows you to critique the effectiveness of the graphics and how they might be redesigned. Through teacher-led activities, you will also be exposed to how graphic communication has evolved, its current practice, and its societal implications.

PROJECT 2: Data-Driven Graphics—Graphing Maximum and Minimum Temperatures

In this project, you will create data-driven visualizations (line graphs and bar charts) using climatic data. These visualizations compare and contrast the climatic conditions in three different parts of the country. This information visualization is designed to help make decisions concerning the heating and cooling costs in different parts of the country. You will create these visualizations by transforming numeric data found on the Web—interpreting how the data is organized into variables, and creating different types of data-driven charts and graphs that serve different communication needs.

PROJECT 3: Data-Driven Graphics—Graphing Degree-Day Data

In this project, you will create data-driven visualizations (line graphs and bar charts) using climatic data. These visualizations build on the work done in Project 2 of this unit by creating charts and graphs that can be used to compare the heating and cooling costs in three different parts of the country. You will create these visualizations by transforming numeric data found on the Web—interpreting how the data is organized into variables, and creating different types of data-driven charts and graphs that serve different communication needs.

PROJECT 4: Conceptual Graphics—The Value of Insulation

In this project, you will create a static, 2D concept-driven visualization about how building insulation works. This will involve exploring different ways of representing the key conceptual elements of the visualization.

Intermediate Projects

PROJECT 5: Animating Insulation Principles

You will create a 2D animation of how insulation retards the flow of heat energy. Starting with the 2D static graphics created as part of the introductory level projects, you will represent the changes in the insulation system over time using 2D animation techniques.

PROJECT 6: Multimedia Presentation of Insulation Properties

You will build on the animation created in Project 5 by demonstrating the heating costs related to different levels of insulation. These animations will explore alternate ways of representing the flow of heat across the insulation barrier. In addition, these animations will be merged with graphs showing how insulation values and the difference in inside and outside temperatures translate into cost to heat a space.

Advanced Projects

You will complete an independent project through the use of visualization tools by researching a new topic dealing with insulation technology or climatic data, or by expanding on topics covered in this unit. The objective of the advanced level projects is for you to further your skills in integrating research, problem solving through the design brief approach, and presentation. It is up to your teacher to work with you to negotiate the topic, time allocated to the project, and design constraints.

IV. Unit Resources

The Resource index at the end of this book contains a listing of all resources associated with this unit. Included are relevant Web sites, books, and other publications. The Glossary provides definitions for all key terms listed in each project.

PROJECT 1
Interpreting Graphics

I. Project Lesson Plan

1. Project Description

This project will provide you with the opportunity to explore how graphics are currently being used to convey technical and scientific information. It also allows you to rehearse the analytic process of decomposing graphics into their component parts and identifying the role of each component. This analytic process also allows you to critique the effectiveness of the graphics and how they might be redesigned. Through teacher-led activities, you will also be exposed to how graphic communication has evolved, its current practice, and its societal implications.

2. Learning Objectives

- Appreciate the historical perspective of graphics as a communications technology by showing how graphics have been used as a communication tool in different periods of history. This exploration includes how graphics were used in early languages and alphabets and later as a tool in technological design. Finally, understand how graphics have been used in scientific exploration and in communicating scientific, technical, and safety-related information to the larger public.

- Follow a standard design process for graphic communication of technical and scientific information. This will entail understanding how scientific/technical information can be decomposed and organized so that it can be graphically represented.

- Be aware of both inadvertent and purposeful misrepresentation of information with graphics. In particular, realize that all graphic communications have a purpose and an audience and that understanding this is central to evaluating the quality and effectiveness of a graphic communication.

3. What You Need

No special materials or software are required. You can use newspapers, magazines, the Internet, and other sources for the graphics.

II. Background Information

1. Historical Perspective

The creation of graphic shapes and forms on surfaces has been a human activity for the past 20,000 years. Early technologies using the application of **pigments** to surfaces such as stone or ceramic have changed little over the years (Figure 1-1 and Figure 1-2). Their purpose has also been a constant: to communicate to others. Whether the communication is for religious, social, aesthetic, or more practical purposes, it makes use of humans' innate ability to process and comprehend visual information.

Figure 1-1 American Indian Cave Painting Primitive creation of graphics used for communication.

(Veni/istockphoto.com)

Figure 1-2 Ancient Mexican Art A ceramic pot painted with the image of Tlaloc, the god of rain and fertility in ancient Teotihuacan culture, in the Templo Mayor Museum in Mexico City.

(Ronaldo Schemidt/AFP/Getty Images)

As Western and Eastern civilizations evolved, more sophisticated systems of graphic communication developed. By the fourth millennium B.C., the Egyptians had begun to develop their written system of **hieroglyphics**, based on pictorial symbols (Figure 1-3). This graphic symbol system continued to develop until roughly 500 B.C. with almost 700 different symbols in use.

Parallel with the development of Egyptian hieroglyphics were a number of other writing systems, including another hieroglyphic system developed by the Mayan culture in the Americas and **cuneiform** writing in the ancient Middle East. This wedge-shaped symbol system was in widespread use throughout the Middle East in the last three millennia B.C.

In the Mediterranean region, the Phoenician **alphabet** developed in the eleventh century B.C. and was the probable basis for the Greek alphabet. The Greek alphabet, in turn, was the basis for both the Latin alphabet and modern European (and American) alphabets. All of these alphabets and hieroglyphic systems have

Figure 1-3 Egyptian Hieroglyphics Ancient graphic symbol system used for communication.

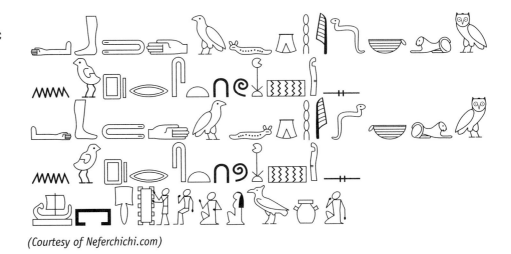

(Courtesy of Neferchichi.com)

the common strategy of using graphic symbols to convey information to others. Alphabets were a development over ancient cave paintings in that they provided a system of symbols and indicated how they could be used in combinations to convey sophisticated and complex information precisely.

In parallel with the development of alphabets was the development of other symbol systems for use in mathematics, science, and technology. These communication systems often used combinations of alphabet characters, words, specialized characters, and other graphic techniques. Similarly, graphic techniques developed standardized methods of representing two- and three-dimensional forms.

During the **Renaissance, pictorial drawing** techniques were developed that allowed artists, scientists, and technologists to create images that reflected the world as they saw it and how it might be (Figure 1-4). From these techniques, the beginnings of the **engineering** profession provided more systematic methods of representing large-scale civil engineering projects, often battlements and other large structures used in defense and war (Figure 1-5). These types of drawings were often more like **diagrams**—representing objects not as they would be seen with the human eye, but in ways that help describe their function or construction.

Moving into the scientific and industrial revolutions, other standardized forms of representation began to evolve. Standardized systems were developed both to

Figure 1-4 Leonardo Da Vinci's Aerial Screw Da Vinci, along with many other early artists, scientists, and technologists, used graphics to depict the world as they saw it and how they imagined it could be.

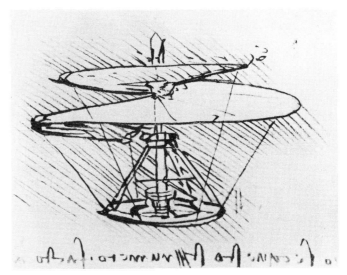

(Detail of a design for a flying machine, c.1488 [pen and ink on paper], Vinci, Leonardo da (1452-1519) / Bibliotheque de l'Institut de France, Paris, France / The Bridgeman Art Library)

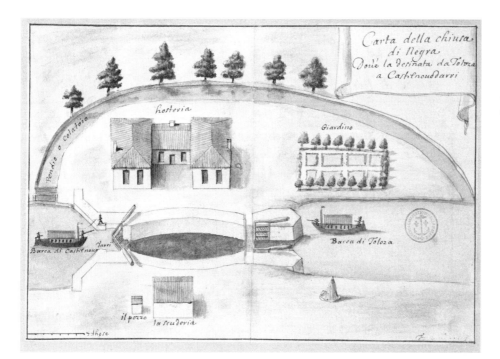

(Drawing of the lock at Negra on the Canal du Midi (w/c on paper), Andreossy, Francois (1633-88) / Service Historique de la Marine, Vincennes, France / Archives Charmet / The Bridgeman Art Library)

Figure 1-5 The lock at Negra on the Canal due Midi This map of the lock at Negra on the Canal du Midi is an example of early pictorial drawing techniques that opened the doors to the engineering profession.

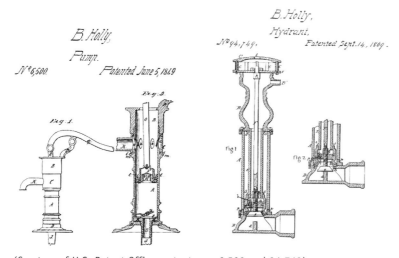

(Courtesy of U.S. Patent Office; patents no. 6,500 and 94,749)

Figure 1-6 Holly Fire Protection and Water System In 1863, Birdsill Holly created these engineering drawings to patent and produce his ingenious Fire Protection and Water System.

represent the designs of physical objects (Figure 1-6) and more conceptual ideas in science and economics. During the eighteenth century, graphs began being used as a vehicle for communicating numerical data, whether it was concerning economic or scientific information. William Playfair used **graphs** to discuss the national debt of England (Figure 1-7). Johann Lambert used graphs extensively in his scientific work, in our example to show solar warming throughout the year at different latitudes (Figure 1-8). A landmark event in public health occurred in 1854, when Dr. John Snow plotted the location of deaths from cholera in central London and used the graphic representation to tie the deaths to contaminated water from the Broad Street water pump (marked with an "X" on the drawing in Figure 1-9). By removing the pump, an epidemic that had killed 500 people was stopped.

As we enter the twenty-first century, industry and commerce continue to integrate across countries and cultures. Governments and businesses have discovered the power of graphics to bridge language barriers. International systems for driving, accommodations, and warnings have been developed and implemented

Figure 1-7 William Playfair's Graph William Playfair used graphing techniques to explain economic trends, seen through this graph he created to present the British national debt from 1699 to 1800.

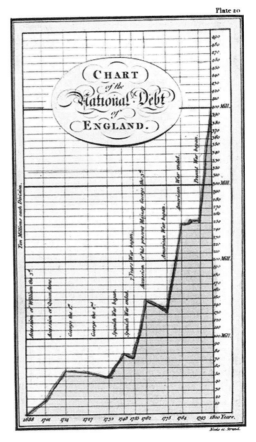

(William Playfair, The Commercial and Political Atlas: Representing, by Copper-Plate Charts, the Progress of the Commerce, Revenues, Expenditure, and Debts of England, during the Whole of the Eighteenth Century, 3rd ed. (London: Wallis, 1801), plate 20)

Figure 1-8 Johann Lambert's Graph of Solar Warming Johann Lambert used graphs to explain scientific concepts, such as this graph he used to display solar warming throughout the year at different latitudes.

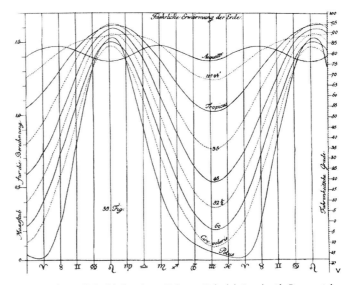

(From Johann Heinrich Lambert, Johann Heinrich Lambert's Pyrometrie; oder, Vom Maasse des Feuers und die Wärme mit acht Kupfertafeln [Berlin: Haude & Spener, 1779], figure 35.)

(From Gilbert, E.W. *Cholera deaths map of central London, 1854. Cited in: Pioneer map of health and disease in England.* Geographical Journal, *124, 172–183, 1958, Blackwell Publishing)*

Figure 1-9 John Snow's Cholera Fatality Map In 1854, Dr. John Snow created this map of cholera deaths in Central London to solve and in turn bring an end to the epidemic.

(Courtesy of U.S. Department of Transportation)

Figure 1-10 United Nations Transport Symbol for Corrosive Substances This international symbol indicating the transport of dangerous goods is one example of how graphics can be used to bridge language barriers between countries.

in numerous countries (Figure 1-10). However, different professions have also developed their own specialized graphics or "traditional uses" of standard graphics to fit the needs of their own work. The types of graphics used in a business presentation may be very different than the graphics used in a medical research journal that, in turn, may be very different than the graphics used on a control panel in a nuclear power plant.

One technological change that has affected all areas of graphics has been the rapid rise in computing power over the past 25 years. While **mainframe computers** were creating simple graphics in the 1950s, their costliness meant that the graphics being produced were for highly critical government and industrial applications, usually defense-related. Up until the 1970s, limitations in displaying graphics meant that screen displays were one-color line drawings, and printouts had to be generated by arranging text symbols on the paper. The lack of computing power during this time also meant that only a limited amount of information could be processed.

The revolution in **scientific and technical visualization** came in late 1970s and early 1980s with the rapid increase in computing power. Greater computing power increased the possibilities for how graphics could be displayed, both on paper and on the computer screen. More computing power also meant that much larger quantities of data could be processed and generated. The rapid increase in computing power brought both the possibility of greater graphics sophistication and the need for ways of managing all of the data being produced. It was this combination that led to individuals exploring the possibilities of how computer graphics could be used to display, or visualize, large quantities of numerical data.

Parallel to these developments in the scientific and technical community were artists and designers exploring the uses of computers for digital graphic production. In the 1980s, **personal computers (PCs)** running the Macintosh operating system paved the way for the new field of desktop publishing and other two-dimensional computer graphics, while larger workstations running the UNIX operating system provided a platform for increasingly sophisticated three-dimensional graphics and animation software. By the turn of the century, workstation and PC technology has merged, with desktop publishing, Web graphics, animation, 3D modeling, and digital video production all available on affordable computers.

Along with the merger of PC and workstation technology has been the union of art and design graphics and scientific and technical visualization. Now many of the same graphic tools can be used to produce graphics that are both functional and aesthetic. That is, the same tools and techniques used to create artistic work that might be seen as part of an animated movie may also be used to help physicists understand subatomic particles. Individuals having special training with these graphic tools are hired to collaborate with the designers and scientists to create these graphic images.

2. Technology and Design

a. Overall Goals

For graphic specialists to create an effective graphic, they have to solve a **design problem**. The problem is a universal one in communications technology: How do you communicate the necessary information to the intended audience with the resources you have at your disposal? To answer this question, the primary goal you need to meet is *communicate graphically the intended information as accurately and as efficiently as possible*. In meeting this goal, the graphic specialist has to work within certain resource constraints. Some of these constraints include:

- Available computer graphics software and manual graphic tools
- Available computer resources
- Available time and people to complete the project
- Ability to use the graphics tools

To successfully complete this project, the graphic specialist needs to:

- Understand who the audience is for the information.
- Understand the specifics of the task. What is to be communicated and for what purpose? Also, how is the information going to be communicated?

- Master the tools—manual and computer-based—needed to create the graphics.
- Master the techniques and processes used to code information graphically and to design its layout, or arrangement.

b. Misrepresentation with Graphics

An overarching goal for all graphic communications is to avoid misrepresenting the information. **Misrepresentation** can come from any of the items listed previously that are needed for successful completion of the project. A lack of personal understanding of the information being communicated is the easiest way to misrepresent the information. Similarly, a lack of understanding of the audience can lead to designing graphics that cannot be properly understood because of the audience members' age, educational level, or cultural background. A lack of mastery of the tools and techniques of graphic communication can mean that the creator of the graphic simply lacks the capabilities to communicate as effectively as might be possible.

The types of misrepresentations to be most careful of are those that can be easily avoided. Purposeful misrepresentation of information is rarely a worthwhile goal when communicating scientific and technical information. It is no more ethically sound than misrepresenting information in written words or speech. Misrepresentation can also come about because the graphic technician did not take the time and effort necessary to execute the best possible design. While all projects have resource limitations, all graphic designers should put forth their best efforts to communicate effectively.

c. Understanding What Is to Be Communicated

Understanding the **audience**, what their goals are for using this information, and what your goals are for communicating, are important beginning steps. To achieve successful communication, you as the graphic specialist (**graphician**) must have a firm understanding about the information you are going to convey. For the activities in this book, we are working with scientific and technical information. To make decisions about displaying this type of data, you can ask a few basic questions.

What area of science or technology best describes the source of this data? In many cases, this data comes from an area that can be described as a **system**. Systems are researched by scientists and technologists so that they can understand how systems "behave" under certain circumstances. For example, scientists and technologists have described a system that explains the relationship between temperature and pressure for liquids; if you heat water in a closed vessel, its pressure will rise. This system could be described graphically in a number of different ways. For example, a graphic could be produced that shows a representation of a heat source, a vessel containing the water, and a gauge measuring pressure (Figure 1-11A). Another way to represent the same system would be to conduct an experiment where you heat the water to various temperatures and measure pressure at each stage. These data values could then be put in a graph showing the relationship of temperature and pressure (Figure 1-11B).

d. Choosing the Best Type of Graphic

The graphic in Figure 1-11A is called a **concept-driven** graphic because it represents the system as an idea, or concept, without the use of specific data values. In the second type of graphic (Figure 1-11B), a **data-driven** graphic, specific information has been generated to show how the system behaves. This data-driven graphic can display the results of tests run on the system, information gathered while the system is in operation, or data generated by mathematical models that theorize how this system might behave. Both of

Figure 1-11 Graphical Representation of Systems Analysis These graphics represent the same scientific concept; however, the audience must be known first to determine which graphic will convey this concept more effectively.

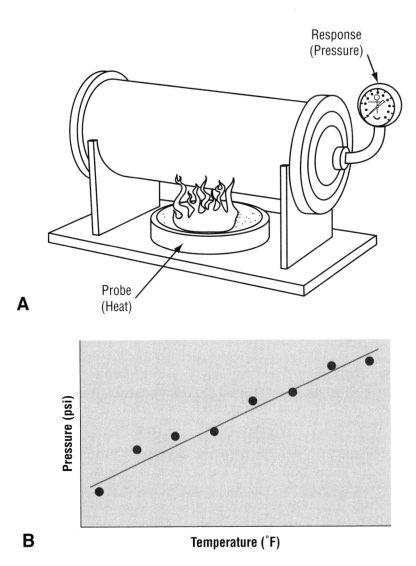

Response (Pressure)

Probe (Heat)

A

Pressure (psi)

Temperature (˚F)

B

these graphics are legitimate representations of the system. Which one is better depends on what the graphician wants to communicate to the audience and the needs of the audience. In many cases, concept-driven and data-driven graphics are used together to give a broader view of a system.

Whether you are creating a concept-driven or a data-driven graphic, you need to make a "mapping" between the information being presented and the graphic being produced. With the previous concept-driven graphic, heat is being represented by a graphic element that looks like a flame. Similarly, elements represent the vessel holding the water, the water itself, and the pressure gauge. With most concept-driven graphics, you have to decide how to break the system up into component parts, what colors and shapes you will use to represent these parts, and, for elements that exist in the physical world, how realistic to make them.

With a data-driven graphic, more established rules are used to represent the different elements of the graphic. For example, data collected from testing or experimentation can usually be defined as **variables**. Depending on the nature of these variables, a type of graph or chart would be selected. Choices of color and shape are made to map these variables on the graph. More information and examples of how to map information to graphics will be given in the activities that follow.

Coming up with a design for your graphic will also depend on the tools that you have to create the graphic. An early decision you often make is whether your graphic is best represented as a **two-dimensional** or **three-dimensional** image. The addition of a third dimension should not be taken lightly; it adds complexity to the production of any graphic. However, if you have specific

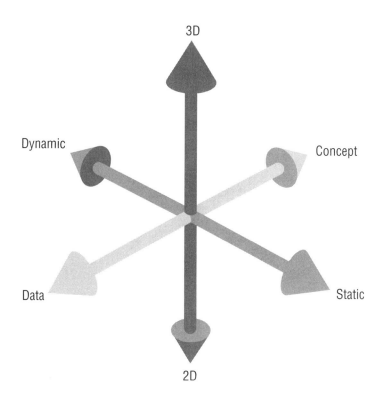

information that can be best represented with the addition of this dimension, then it might be a choice to consider. Similarly, you need to decide whether the graphic should be **dynamic**—that is, does it change over time—or **static**, that is, unchanged. If part of what you want to communicate about your system is how it changes over time, then dynamic elements can be helpful. Whether dynamics or a third dimension are included will also depend on the capability of your computer software, manual tools, and your own skill with these tools.

In summary, your analysis will lead you to decide whether your graphic will be (Figure 1-12):

- Concept-driven or data-driven
- Two-dimensional or three-dimensional
- Static or dynamic

The graphic can be any combination of these factors depending on the specific communication needs and the resources you have at your disposal. All things considered equal, a three-dimensional dynamic graphic will be more complex and take more planning and effort than a static two-dimensional graphic. In many cases, you may decide that multiple graphics are needed to convey the necessary information and that your graphics will use different combinations of these factors (Figure 1-13).

4. Technology and Society

Technical and scientific visualization is a technological problem in part because it is solving the problem of communicating information, either to oneself or to others. In doing so, the goal will always be to communicate this information as accurately as possible within the constraints of time, materials, tools, and knowledge. Failure in communication can have many consequences. It can simply mean that individuals have to expend more time and effort to understand the information being presented. It can also lead to misunderstanding of the information. In both these cases, the result may mean lost time, lost money, or in the most extreme cases, injury or loss of life.

Figure 1-13 Final Composition of a Visualization The final composition of a visualization may include multiple graphical elements to convey the necessary information.

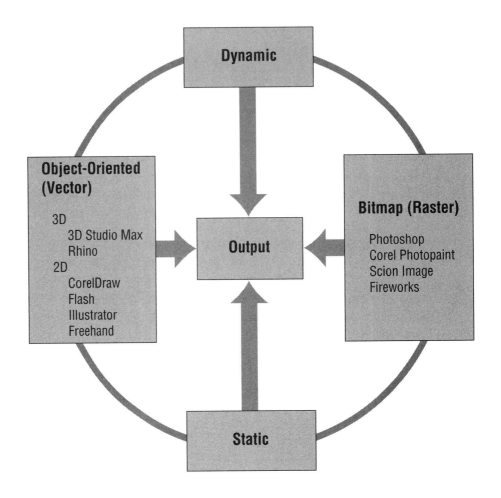

Key Terms

pigments
hieroglyphics
cuneiform
alphabet
Renaissance
pictorial drawing
engineering
diagrams
graphs

mainframe computers
scientific and technical
visualization
personal computers (PCs)
design problem
misrepresentation
audience
graphician

system
concept-driven
data-driven
variables
two-dimensional
three-dimensional
dynamic
static

III. Student Materials

Use Project 1 Design Brief: Graphical Analysis to complete the project.

UNIT 1: PROJECT 1   DESIGN BRIEF: GRAPHICAL ANALYSIS

You work for a science and technology textbook publisher. It is your job to research and acquire graphics used for reinforcing and illustrating the concepts presented in textbooks your publisher produces. For most projects, you must assess graphics you acquire or produce in-house with the following tasks:

- Identify the audience(s) for graphics.
- Decompose the graphic into its component parts (e.g., independent and dependent variables, concept elements).
- Try and understand what information the graphic is trying to convey.

Collect example graphics meant to communicate information about a particular topic.

- Your teacher will tell you how many examples you need to collect.
- You can use newspapers, magazines, the Internet, and other sources for your graphics.
- Reproduce them in color when necessary.
- Make sure that you include complete citations for all of your examples.

Be prepared to report to your teacher with your analysis for each graphic.

PROJECT 2
Data-Driven Graphics: Graphing Maximum and Minimum Temperatures

I. Project Lesson Plan

1. Project Description

In this project, you will create data-driven visualizations (line graphs and bar charts) using climatic data. These visualizations compare and contrast the climatic conditions in three different parts of the country. This information visualization is designed to help make decisions concerning the heating and cooling costs in different parts of the country. You will create these visualizations by transforming numeric data found on the Web—interpreting how the data is organized into variables, and creating different types of data-driven charts and graphs that serve different communication needs.

2. Learning Objectives

- You will learn about the role of climatic conditions on the location and design of a manufacturing facility by exploring the impact weather conditions play on the effectiveness of various transportation systems and on the operation of the manufacturing facility. The focus will be the heating and cooling costs of the manufacturing facility and the role average outside temperatures and the insulation design play in these costs.

- You will access and repurpose climatic data maintained by the National Oceanic and Atmospheric Administration (NOAA). This process will involve understanding how the data is displayed in tabular format on the Web and how it can be captured and imported into a spreadsheet (Excel).

- You will use spreadsheet software (Excel) to generate line and bar charts from the climatic data. This will involve identifying the variables to be displayed, the best type of chart for the type of data, and the pros and cons of different types of displays.

3. What You Need

- Student design brief
- Internet access
- Web browser
- Spreadsheet software, such as MS Excel.

Optional

- PowerPoint presentation software
- Printer
- Word processor software for viewing the raw data

II. Background Information

1. Raw Data Source

The National Oceanic and Atmospheric Administration of the U.S. Department of Commerce is responsible for collecting and managing weather data from across the United States and the rest of the world. Through their National Data Centers, weather (climatic) data is available for free or nominal cost (see the *Resources Index* for URLs). One collection of data—called a *product*—is the minimum and maximum temperature for cities across the United States averaged over the past 30 years. For **climatic** information on cities of interest, visit the Web sites listed in the Resources Index.

The NOAA Web site has a set of tables called **climatological normals**, including normal daily maximum and minimum temperatures. In these tables, the average (normal) minimum and maximum temperature for each month, calculated from 30 years of data, is listed for major weather stations across the country. If you look at this table of data, you will see that the first few rows of text give basic information concerning the contents of the table (e.g., normal daily maximum temperatures and degrees Fahrenheit). The next important row labels the columns of data. The first column, labeled "NORMALS 1971–1990," sits over the weather station locations. The next, labeled "YRS," indicates the number of years of data that have been used in the calculations for each weather station (30 years in almost all cases). The next 12 columns label the average for each month of the year. The last column is the average for the whole year. Therefore, each row provides temperature data for a single weather station.

2. Importing Data into Excel

Our first step will be to transfer this data from the Web page to a **spreadsheet** in Excel where we can more easily manipulate and visualize the data. Start by clicking on the normal daily maximum temperature link. The data is displayed in table form and can be easily saved and imported into Excel. Using Internet Explorer or another Web browser, bring up the Web page and go to File > Save As … and save the page in Plain Text format. Now go to Excel. The page we saved from Explorer is what is called a raw or **ASCII** text file. It has to be imported into Excel and saved in its native format. Start by going to File > Open and choosing the text file you previously saved. The Text Import Wizard will come up and look something like this (Figure 1-14):

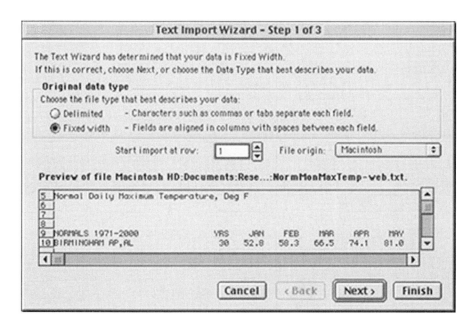

Figure 1-14 Text Import Wizard – Step 1 of 3 This dialogue box allows you to import properly formatted text into an Excel spreadsheet.

To import a text file such as the one we saved from the browser into Excel, you have to define how the text will be divided up into the different cells. Each row of text goes into a row of cells in Excel, so this is pretty straightforward. The more difficult task is to define how the text is divided between the different cells in a row—the columns. The first of two primary methods is to use a standard character, such as a comma or tab, to indicate when to move to the next cell in a row. This is called delimited text. The other method is to place text in columns by a count of the number of characters from the left edge. This is called fixed width and is the method we will use.

When you proceed to the next screen after selecting Fixed width (Figure 1-15), the Wizard looks at the spacing between the text and tries to estimate the best location for creating columns. For this file, you can move the first break line for the first column over to the thirty-second character. This will ensure that all of the weather station name text ends up in the first column. Each break line after this should be placed right after the last character of the column data. For example, the YRS column ends at character 35, the JAN column at 42, and so on.

Figure 1-15 Text Import Wizard – Step 2 of 3 Once done, go to the next page of the wizard (Figure 1-16).

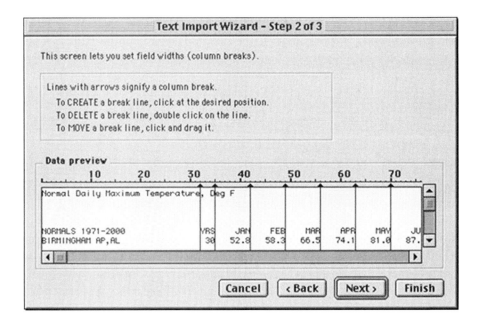

Figure 1-16 Text Import Wizard – Step 3 of 3 In this screen, the only task is to set the format for the first column to text. The remaining columns contain numeric data and can be left in the general format.

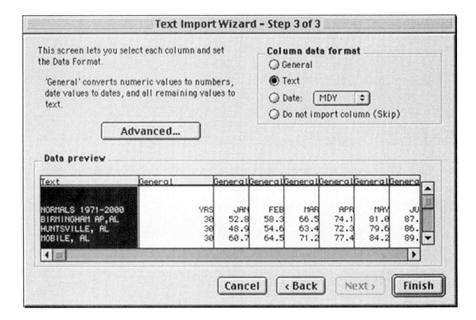

3. Organizing Your Data

Once you leave the text wizard, the text is placed in the spreadsheet cells as you specified in the wizard (Figure 1-17). The first couple of rows containing general information about the table have been poorly broken up between the rows and columns, but the temperature data (the critical content!) has been properly divided up between the cells. Highlight the cells containing temperature data and pick Format > Cells.... From this dialogue box, you can format the cells to be numeric with one decimal place of precision. You can also clean up the text in the first few rows.

With the table of data properly placed in Excel cells, you can now select your weather stations of interest by copying the appropriate rows to a new page in the Excel workbook (Figure 1-18). You may also want to start labeling the pages of the workbook to keep track of the information.

On this page, the label for the first column was renamed "City" to be more descriptive. Also, the column containing the number of years the data was collected was deleted because it won't be used any more.

4. Visualizing Your Data

a. Deciding on the Type of Graph or Chart

So now we have the average maximum daily temperature for our three candidate cities. Though you can look at the numbers on the spreadsheet and try and make some "sense" of the data, graphing it would give you a better overall view of the information. First, you must decide what would be the most appropriate type of graph. To do so, you must start by identifying what are the dependent and independent variables of the graph. In this case, there are two independent variables: the city and the month of the year. As is typical for a table like this, the labels for these variables are in the far left column and top row. The values for the dependent variable, temperature,

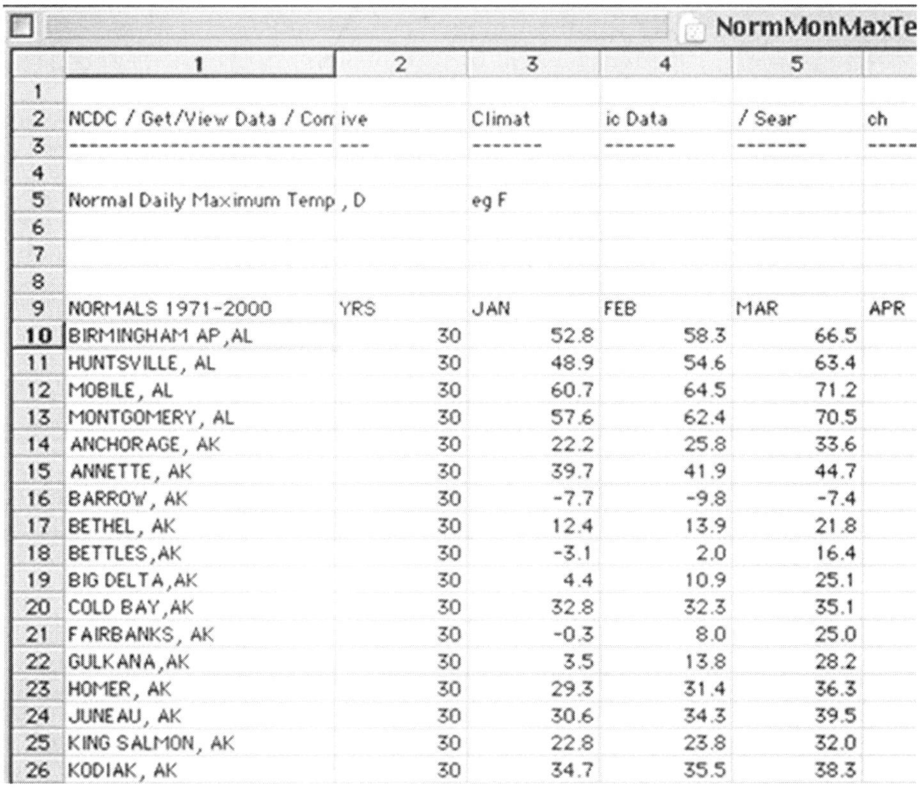

		1	2	3	4	5	
1							
2	NCDC / Get/View Data / Corr ive		Climat	ic Data	/ Sear	ch	
3	------------------------- ---		--------	--------	--------	-----	
4							
5	Normal Daily Maximum Temp , D		eg F				
6							
7							
8							
9	NORMALS 1971-2000	YRS	JAN	FEB	MAR	APR	
10	BIRMINGHAM AP,AL	30	52.8	58.3	66.5		
11	HUNTSVILLE, AL	30	48.9	54.6	63.4		
12	MOBILE, AL	30	60.7	64.5	71.2		
13	MONTGOMERY, AL	30	57.6	62.4	70.5		
14	ANCHORAGE, AK	30	22.2	25.8	33.6		
15	ANNETTE, AK	30	39.7	41.9	44.7		
16	BARROW, AK	30	-7.7	-9.8	-7.4		
17	BETHEL, AK	30	12.4	13.9	21.8		
18	BETTLES,AK	30	-3.1	2.0	16.4		
19	BIG DELTA,AK	30	4.4	10.9	25.1		
20	COLD BAY,AK	30	32.8	32.3	35.1		
21	FAIRBANKS, AK	30	-0.3	8.0	25.0		
22	GULKANA,AK	30	3.5	13.8	28.2		
23	HOMER, AK	30	29.3	31.4	36.3		
24	JUNEAU, AK	30	30.6	34.3	39.5		
25	KING SALMON, AK	30	22.8	23.8	32.0		
26	KODIAK, AK	30	34.7	35.5	38.3		

NormMonMaxTe

Figure 1-17 Normal Daily Maximum Temperature, All Cities This spreadsheet displays the information imported into Excel through the Text Import Wizard with the cities and temperatures broken up into the appropriate rows and columns.

Figure 1-18 Normal Daily Maximum Temperature, Three Cities This spreadsheet displays the normal daily maximum temperature of the three cities of interest.

	1	2	3	4	5	6
1						
2	Normal Daily Maximum Temperature, Deg F					
3						
4	City	JAN	FEB	MAR	APR	MAY
5	SAN DIEGO, CA	65.8	66.3	66.3	68.7	
6	CHICAGO, IL	29.6	34.7	46.1	58.0	
7	BALTIMORE, MD	41.2	44.8	53.9	64.5	
8						
9						
10						
11						
12						

MaxTemp-All \ **MaxTemp-3cities**

fill in the core of the table. Each independent variable has values called levels. For the variable *City*, the levels are San Diego, Chicago, and Baltimore. For *Month*, the levels are Jan, Feb, Mar, and so forth. For each level of each independent variable, there is a unique dependent variable value. In Baltimore in March, the average maximum daily temperature is 53.9 degrees Fahrenheit.

Independent variables can either be nominal (names), ordinal (ordered, but not using a uniform scale), or scalar. Cities is a nominal variable while months could be considered to be either ordinal or scalar. Though there are not an equal number of days in each month, you can treat each month as though it were of equal length for a graph such as this. The independent variable is typically placed on the horizontal axis of the graph. When there is more than one independent variable, then usually the one with the highest number of levels is placed there (months in this case). The dependent variable, temperature, is placed on the vertical scale.

Which type of graph you choose will depend on both on the type and number of variables you have and how you would like to compare them. Using *Month* as the independent variable, you have a choice of either a line graph or a bar chart. Line graphs allow you to look at the angle (or slope) the line makes between dependent variable values to visually determine the rate in which the value is increasing or decreasing. It is possible to make sense of the line slope only if the increments are uniform between each independent variable level. Therefore, the independent variable value should be scalar for us to make a line graph. Let's treat *Month* as a scalar variable and create a line graph for the first city, San Diego.

b. Creating a Line Chart: One Independent Variable

Highlight the row containing the column labels and the row containing the San Diego temperatures up through December. Now choose the Chart Wizard icon on the toolbar (Figure 1-19).

You will be creating a line chart and, given that you have only 12 points, it would be appropriate to indicate each data point with a shape mark (e.g., a diamond).

In Step 2, the Wizard has identified the first row as the labels for the independent variable level, the first column for the second independent variable, *City*, and the remaining values for the dependent variable (Figure 1-20). In the language of the Chart Wizard, a series is the data relating to a single level of the second independent variable (e.g., the temperatures for San Diego). These are in rows, starting with the second row. In charting the data, the

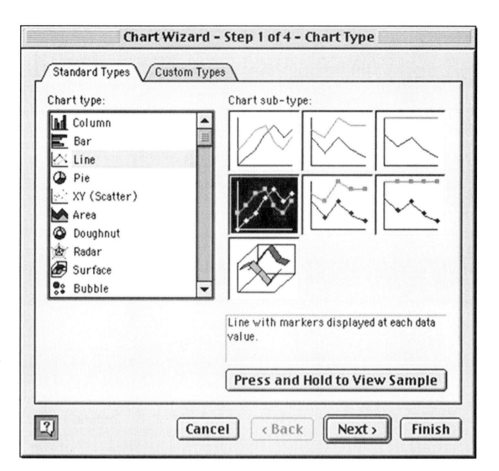

Figure 1-19 Excel Chart Wizard – Step 1 of 4 The Chart Wizard will aid in the creation of an appropriate graph, in this case a line chart with one independent variable.

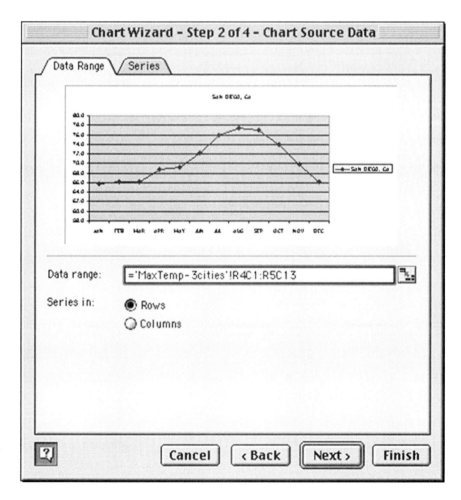

Figure 1-20 Excel Chart Wizard – Step 2 of 4 This step allows the selection of the chart source data.

Figure 1-21 Excel Chart Wizard – Step 3 of 4 This step allows manipulation of the different chart options.

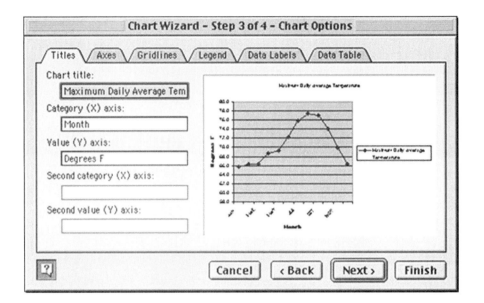

second independent variable is not placed on the horizontal axis, but is listed in a legend. Each level of the second independent variable is coded with a line color and a shape mark. In this case, there is only one level, San Diego. If the second independent variable has only one level, we would probably hide this legend and incorporate the name of this second variable into the title of the graph. However, we will leave the legend, because we will be modifying the graph later. A preview of the graph is seen in this window.

In Step 3 of the wizard (Figure 1-21), a title and variable labels can be entered. Finally, in Step 4, the dialogue box asks whether you would like to place the graph on a new, separate worksheet or on the same worksheet as the data table. It will be easier to modify the table if you keep it on the same worksheet.

When you have finished, your graph should look like the one in Figure 1-22. The only additional change is to increase the visual contrast between the data

Figure 1-22 Finished Excel Line Graph The data and axis values used to create the line graph are highlighted on the spreadsheet page above the graph.

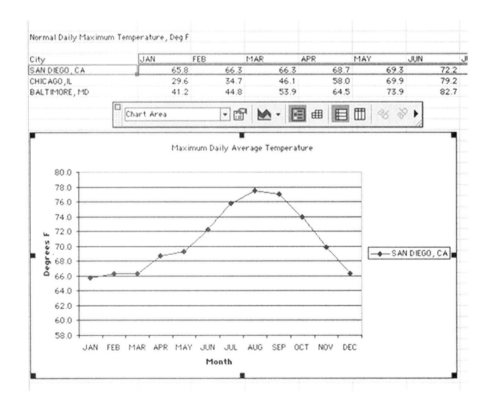

points and the background by turning the background white. This is done by double-clicking on the background gray and choosing None for the background area color. Note that when the chart is highlighted, the source data for the chart is also highlighted. The first independent variable labels are highlighted in magenta (purple), the second independent variable label in green (the series), and the dependent variable values in blue. There is an active link between any of these values and the chart. If you change values or labels in these cells, they will also change in the chart.

You can now repeat this process to create charts for the next two levels of the *City* independent variable. The easiest way to do this is to copy your first chart and then change the *City* variable level (the series). Do this by highlighting the entire chart, choosing Edit > Copy, highlighting the cell where you would like the upper-left corner of the new chart to be, and then choosing Edit > Paste. Now change the series by dragging the green and blue boxes down to the next row (make sure you click right on the colored box edge). You can also do this by going to the menu Chart > Source Data and redefining the name and values of the series.

c. Creating a Line Chart: Two Independent Variables

There are now three charts, each showing a different series of the second independent variable, *City*. Inspect each chart by itself. Where are the areas of the line that are close to horizontal, and where is the line closer to vertical or something in between? Not surprisingly, there are not too many flat (near horizontal) areas, but a gradual slope upward toward higher average temperatures until mid-year, when it starts to slope down. However, each line has its own distinct shape. San Diego has a fairly flat area until April, when it gradually climbs to August. The second half of the year is a much more consistent decline. Chicago and Baltimore have much more uniform curves to midyear and then to the end of the year.

What about comparisons between the different levels of the *City* variable? You can look between the charts to make comparisons, though it is generally harder than making comparisons within a single chart. One thing you might notice is that Chicago and Baltimore may have the same general shape, but Chicago's points are offset nearly 10 degrees lower than Baltimore. You might also notice that San Diego's temperatures vary over a much smaller range. This is hard to see in part because the three charts do not use the same vertical scale range for the dependent variable, *Temperature*. By default, Excel resets the axis scale to match the range of data. In Figure 1-22, the range is from 58 to 80 degrees for San Diego, while Baltimore and Chicago have scales that range from about 30 to 90 degrees.

By double-clicking on the vertical axis for the San Diego chart, you can set the axis scales to fixed values that match the other charts (Figure 1-23). For example, you might set the range from 0 to 90 degrees to cover a wide range of U.S. cities.

With these settings on all three charts, new differences appear. Notice how flat the San Diego line is compared to Chicago and Baltimore. The offset of the Baltimore line relative to the Chicago line is a little easier to see now, but it still has the same basic shape of Chicago. If the San Diego chart were going to be seen all by itself, its original scale would have been appropriate because it helps distinguish the monthly differences. However, when comparing it to other cities, it is appropriate to use a common scale.

The next logical step is to merge all three series onto a single line chart so that direct comparisons can be made between the three cities in a single chart (Figure 1-24). Start this process by making a copy of the Baltimore chart. Then copy and paste the other two charts on top of it. That is, select

Figure 1-23 Excel, Format Axis The Format Axis dialogue box allows the manipulation of the graph's axis.

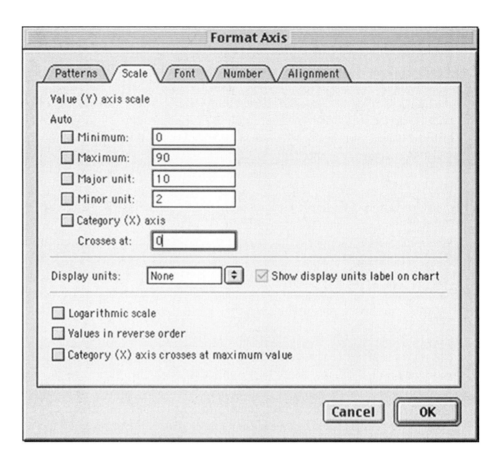

Figure 1-24 Graph of Three Cities' Maximum Daily Average Temperature This graph is the outcome of merging the three individual maximum temperature per city graphs into one, to be used for easier comparison of temperatures between the cities.

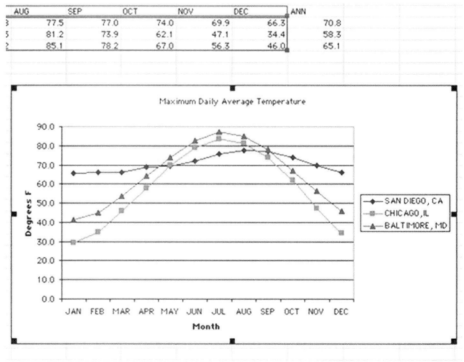

the Chicago chart, pick Edit > Copy, then select the "new" Baltimore chart and pick Edit > Paste. Repeat with the San Diego chart.

Notice that the blue box around the dependent variable values is now around all three levels of the *City* variable. The only additional change made to this chart is changing the color yellow on the Baltimore line to green, which has more contrast with the background. With the three levels of the *City* variable merged on one graph, you can now clearly see the differences in overall line shapes in addition to easier comparisons for any individual month. Note that though Chicago and Baltimore start and end with about a 10-degree offset, they are much closer together for the warmest month of the year, July. San Diego does not peak until August, but at a cooler temperature than the other two. Create a new worksheet and repeat this entire process for the average daily minimum temperatures.

The last thing to do with these two graphs is to indicate on the charts our target temperature for inside the manufacturing facility, 72 degrees Fahrenheit (Figure 1-26).

This can be done by drawing a reference line on the graph with the Drawing tools in Excel (if you do not see the Drawing tools, go to Tools > Customize and pick Drawing from the Toolbars tab). A red line, 1 1/2 points (pts) thick, shows up nicely. You will need to estimate the location of the line on the graph. One of the drawing tools allows you to create a text block to label this reference line.

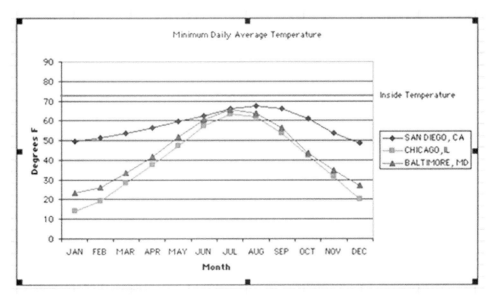

Figure 1-25 Graph of Three Cities' Minimum Daily Average Temperature The target inside temperature is represented with a red line drawn using Excel's drawing tools.

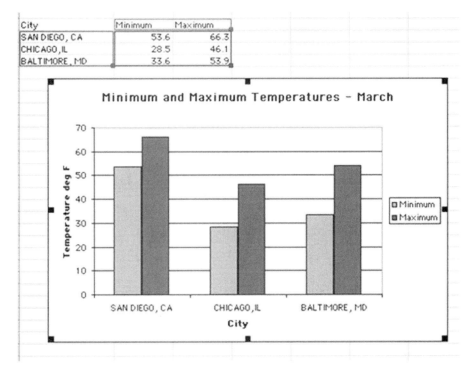

Figure 1-26 Minimum and Maximum Temperatures for March This is a bar chart comparison of the minimum and maximum daily average temperatures of the three cities in March.

d. Creating a Bar Chart: Two Independent Variables

How else can you look at this same data? One alternative would be to focus on the difference between the minimum and maximum temperatures for the three cities on selected months (Figure 1-27). In this case, the two independent variables are *City* and *Minimum Versus Maximum*. Now both of the independent variables are nominal, and a bar chart would be much more appropriate to display the data. Create a new worksheet, gather the appropriate data for the month of March from the other worksheets, and build a new table. Finally, create a bar chart with the *City* variable on the horizontal axis.

Key Terms

climatic	spreadsheet	ASCII
climatological normals		

III. Student Materials

Use Project 2 Design Brief: Planning to Locate Your Biotechnology Company to complete the project.

UNIT 1: PROJECT 2   DESIGN BRIEF: PLANNING TO LOCATE YOUR BIOTECHNOLOGY COMPANY

You are on the facility location team for BioTech, a company looking to build a new manufacturing facility. The three cities that are the top choices are San Diego, Chicago, and Baltimore. These cities are near major transportation routes and otherwise fit the company's criteria. Weather (climatic) data is needed to help calculate the expected heating and cooling costs for the new facility. This is one of the cost factors used to decide where to locate the manufacturing facility. The inside of the manufacturing facility has to be kept at a strict 72 degrees Fahrenheit. The more the outside temperature varies from this reference point, the more it will cost to either heat or cool the facility. You are responsible for presenting information to the management team to help them make their decision. You have only a few minutes for your part of the presentation, and you need to clearly and concisely communicate to them how the three top choices are going to differ in terms of heating and cooling costs.

PROJECT 3
Data-Driven Graphics: Graphing Degree-Day Data

I. Project Lesson Plan

1. Project Description

In this project, you will create data-driven visualizations (line graphs and bar charts) using climatic data. These visualizations build on the work done in Project 2 of this unit by creating charts and graphs that can be used to compare the heating and cooling costs in three different parts of the country. You will create these visualizations by transforming numeric data found on the Web—interpreting how the data is organized into variables and creating different types of data-driven charts and graphs that serve different communication needs.

2. Learning Objectives

- You will learn the role of climatic conditions on the location and design of a manufacturing facility by exploring the impact weather conditions play on the effectiveness of various transportation systems and on the operation of the manufacturing facility. The focus will be the heating and cooling costs of the manufacturing facility and the role average outside temperatures and the insulation design play in these costs.

- You will use spreadsheet software (Excel) to generate line and bar charts from the climatic data. This will involve identifying the variables to be displayed, the best type of chart for the type of data, and the pros and cons of different types of displays.

3. What You Need

- Student design brief
- Internet access
- Web browser
- Excel spreadsheet software

Optional

- PowerPoint presentation software
- Printer
- Word processor software for viewing the raw data

II. Background Information

Graphing Degree-Day Data Tutorial

Ultimately, the charts you produce are serving the goal of comparing heating and cooling costs of the three city candidates. While the charts we have created provide valuable information concerning the differences in temperature, they address our question of interest indirectly. Calculations of energy cost are based on **degree days**. This value represents the number of degrees, each day, the outside temperature varies from the target value. The greater the difference is between the outside temperature and the target inside temperature, the greater the **load** on the heating/cooling system for the building. In this case, the target value is 72 degrees.

Rather than trying to work with both the minimum and maximum temperatures, the average temperature (minimum + maximum / 2) is used. Now set up a new worksheet to generate these values (Figure 1-27). The NOAA Web site has a table of the average temperature if you do not want to calculate it. You can also simplify the calculations by assuming that each month has 30 days.

There are a number of options for creating a chart. One possibility would be to create a bar chart very much like the one created for the minimum and maximum temperatures (Figure 1-28). Cities would again be the **independent variable** on the horizontal axis. *Month* would now be the second independent variable and *Degree Days* the **dependent variable**.

By organizing the independent variables in this manner, you can clearly see the monthly changes in degree day load by city. This chart also clearly shows

R12C2 =ABS((72-R[-7]C)*30)

NormMinMaxTemp-web.xls

	1	2	3	4	5	6	7
1							
2	Normal Daily Minimum Temperature, Deg F						
3							
4	City	JAN	FEB	MAR	APR	MAY	JUN
5	SAN DIEGO, CA	57.8	53.9	60.0	62.5	64.6	67.4
6	CHICAGO IL	22.0	27.0	37.3	47.3	58.7	68.2
7	BALTIMORE, MD	32.3	35.5	43.7	53.2	62.9	71.8
8							
9	Degree (Deg F) Days - Total						
10							
11	City	JAN	FEB	MAR	APR	MAY	JUN
12	SAN DIEGO, CA	426.0	393.0	360.0	282.0	222.0	138.0
13	CHICAGO,IL	1500.0	1353.0	1041.0	726.0	395.0	114.0
14	BALTIMORE, MD	1191.0	1095.0	849.0	564.0	273.0	6.0
15							

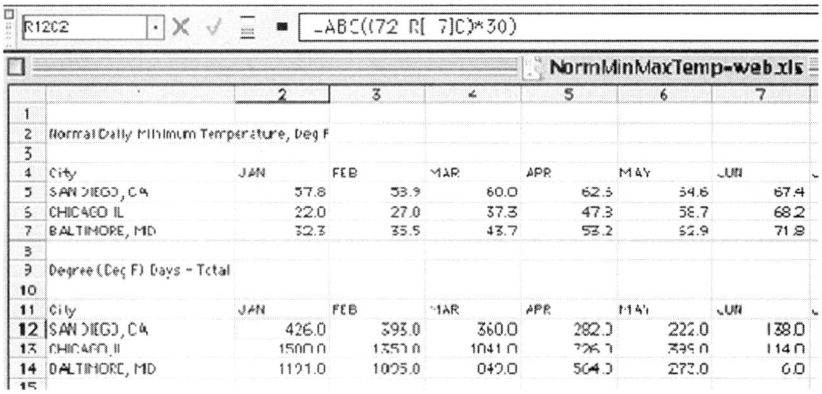

Figure 1-27 Normal Minimum and Maximum Temperatures Spreadsheet Degree-day data is calculated by month for each city.

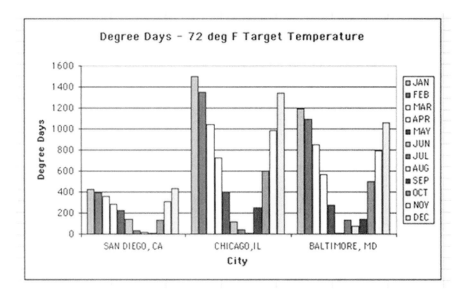

Figure 1-28 Bar Chart of Degree Days by City This graph displays the monthly changes in degree-day load by city.

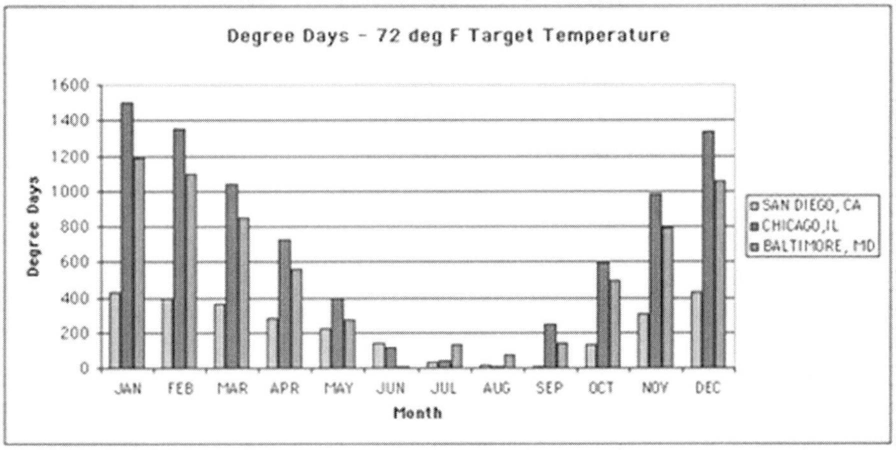

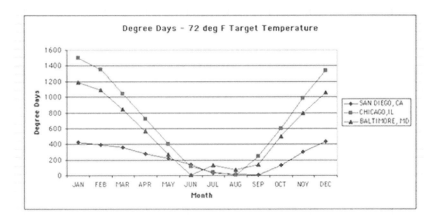

Figure 1-29 Bar Chart of Degree Days by Month

Figure 1-30 Line Chart of Degree Days by Month

the *V* shape the degree-day load goes through over the course of the year. However, San Diego clearly has a shallower *V* than the others. The disadvantage of this chart is the number of levels of the second independent variable. **Color coding** 12 different months makes for a visually "busy" chart.

An alternative would be to switch the two independent variables. This can be done by having the **series** defined by rows in the Chart Wizard (Figure 1-29).

Now the chart has only three levels for the second independent variable. The overall *V* shape is still distinct, though it is more difficult to see the differences in each city's curve shape. However, you can easily compare the difference in degree days between cities for a particular month.

Still another alternative would be to turn this bar chart into a line chart (Figure 1-30).

Here, each city's curve is visible, and the difference in curve shape is readily apparent. Just as you could with the first bar chart, you can visually estimate the area under the curve for each city to estimate the degree-day total for the year. One disadvantage of this display is that it becomes difficult to read individual values when the lines cross or points are close together.

Why do you need these charts? Why not just present the values of the degree-day totals for each city? Understanding the heating and cooling costs and integrating this information into the decision about the city in which to locate the manufacturing facility may be more complex than simply looking at these totals. For instance, changes in heating costs over the course of the year may be a factor depending on the cycle of cash flow during the year. In the end, probably both charts and tables of values will be of use.

degree days   independent variable   color coding

load   dependent variable

III Student Materials

Use Project 3 Design Brief: Planning to Locate Your Biotechnology Company to complete the project.

UNIT 1: PROJECT 3   DESIGN BRIEF: PLANNING TO LOCATE YOUR BIOTECHNOLOGY COMPANY

You are on the facility location team for BioTech, a company looking to build a new manufacturing facility. The three cities that are the top choices are San Diego, Chicago, and Baltimore. These cities are near major transportation routes and otherwise fit the company's criteria. Weather (climatic) data is needed to help calculate the expected heating and cooling costs for the new facility. This is one of the cost factors used to decide where to locate the manufacturing facility. The inside of the manufacturing facility has to be kept at a strict 72 degrees Fahrenheit. The more the outside temperature varies from this reference point, the more it will cost to either heat or cool the facility. You are responsible for presenting information to the management team to help them make their decision. You have only a few minutes for your part of the presentation, and you need to clearly and concisely communicate to them how the three top choices are going to differ in terms of heating and cooling costs.

PROJECT 4
Conceptual Graphics: The Value of Insulation

I. Project Lesson Plan

1. Project Description

In this project, you will create a static, 2D concept-driven visualization about how building insulation works. This will involve exploring different ways of representing the key conceptual elements of the visualization.

2. Learning Objectives

- You will use graphics created with an object-oriented 2D graphics software tool (CorelDraw, Fireworks, Freehand, etc.) to communicate about insulation technology.

- You will create graphics that use color and shape elements—supported by text—to communicate how insulation retards the movement of heat energy from a hotter geographical region to a cooler geographical region.

- You will identify the key concepts and elements to be displayed and investigate how to best code and arrange these elements in a graphic (or graphics).

3. What You Need

- Student design brief
- Internet access
- Web browser
- Object-oriented 2D graphics software (e.g., CorelDraw, Fireworks, or Freehand)

Optional

- PowerPoint presentation software
- Printer
- 2D dynamic software package (e.g., Flash)
- 3D modeling and animation software (e.g., 3D Studio Max)

II. Background Information

1. Planning Your Graphic Communication

In this problem, you are going to be communicating basic principles of **insulation** and insulation technology to decision makers within the biotechnology company. Your goal is to help the company make decisions about what kind of insulation technology should be used and assist in the calculations of the cost trade-offs of more insulation versus ongoing heating costs. You may represent a construction company submitting a proposal to the biotechnology company for insulating their new facility.

To design the graphics, the key concepts to be communicated need to be identified:

- The purpose of insulation is to retard the movement of heat (heat transfer). This transfer of heat is normally thought of as the movement from a high temperature object or area to a lower temperature object or area. Heat is directly related to the internal energy of both objects/areas involved, according to the First Law of Thermodynamics. For the purpose of this communication, a detailed knowledge of the First Law of Thermodynamics is not needed. However, the concept of heat and the movement of heat (energy) from one location to another is necessary.

- The different materials used in construction of the building that play a role in insulating the building need to be represented, as do any other structural materials that house/hold the insulation.

- One critical quality of insulation is how much heat passes through it in a unit of time. Does time have to be represented and if so, how?

- There are three main types of heat transfer: conduction, convection, and radiation. Conduction is the main cause of heat loss through the building walls and ceiling. Should it be the only one represented?

Let's start with representing the wall of the building. The building envelope consists of the barrier between the inside and the outside, and it typically consists of the walls, ceiling/roof, and ground floor/foundation of the building. For the purpose of this project, we will primarily be concerned with the walls, though the ceiling/roof and ground floor/foundation may very well use a different kind of insulation technology. A wall typically consists of a layer on the outside of the building, a layer on the inside, and the space in-between. The space in-between can hold systems such as plumbing or electrical, but more importantly for this discussion, it also typically holds insulating material. The outer layer and inner layer need to be represented differently because they are made of different materials that serve different purposes (e.g., the outer layer has to be weatherproof while the inner layer does not).

How would we represent this wall? Each element will need to be represented with its own symbol. Symbols need to be differentiated from each other by changes in visual characteristics such as color or shape. You can also take advantage of symbols and colors that target audiences use in their everyday settings.

2. Example of Concept-Driven Graphics

To begin with, the typical way of viewing a wall is to show it in cross-section (Figure 1-31). This is a convention used in the construction industry and should be recognizable to an audience used to construction graphics. To other audiences, you might need to describe what you are seeing. A wall section might look like this:

Here the **fiberglass** insulation is represented using a standard symbol used in the construction industry—a wavy line. The outer sheathing is represented in the darkest color, the **composite board** a slightly lighter gray, and the **drywall** (interior wall) in light gray. However, these symbols may not be self-evident. Also, more information may be needed about the types of sheathing. Text may be appropriate to use in this case (Figure 1-32).

In addition to text labeling, another way to provide context for the different elements of the wall is to show the third dimension. With an oblique projection (Figure 1-33), the cross-section would still be shown as it was in the previous figures, but now the third dimension would extend at an angle to one side. This is an

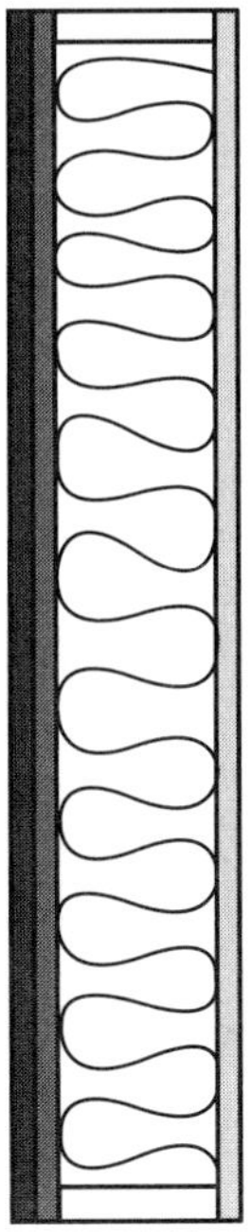

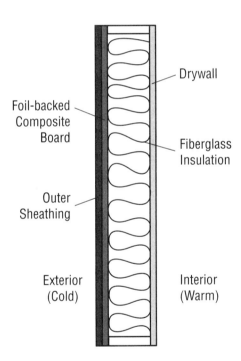

Figure 1-31 Cross-Section of a Wall This cross-section displays the composition of a wall. In this example, the wall has four components: fiberglass insulation, outer sheathing, composite board, and drywall.

Drywall

Foil-backed
Composite
Board

Fiberglass
Insulation

Outer
Sheathing

Exterior
(Cold)

Interior
(Warm)

Figure 1-32 Cross-Section of a Wall with Text The same cross-section with text added distinguishes the wall components with more clarity.

abstracted representation of an actual wall, but it does help show more information. What is the downside of such an approach? Drawing the third dimension does take time, and the illustration now takes up more space. Maybe most importantly in this case, there is now longer equivalent blank space to put additional information on both the exterior and interior side of the wall.

In Figure 1-34, we've returned to the cross-section view and added large shaded regions that help emphasize the interior and exterior spaces. In your representation, you can use colors that are traditionally used in Western culture indicate hot (warm) and cold (cool). In addition, relatively light, low saturation colors are used for this large background area. These colors allow other symbols to be placed on top of them and still be visible. Here, we've used different patterns of lines to represent the temperature difference. Then, two arrows in darker, higher saturation shades of the same colors are placed on top of the backgrounds to indicate the

Figure 1-33 Pictorial of a Cross-Section of a Wall The same cross-section is shown as an oblique pictorial projection. This projection shows the third dimension of the wall.

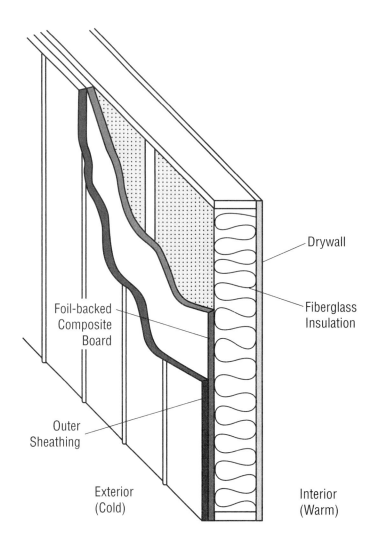

Drywall

Fiberglass Insulation

Foil-backed Composite Board

Outer Sheathing

Exterior (Cold)

Interior (Warm)

Figure 1-34 Cross-Section of a Wall with Coded Regions and Arrows The same cross-section with coded regions added distinguishes the interior and exterior spaces with more clarity; also, the arrows show the direction of heat flow.

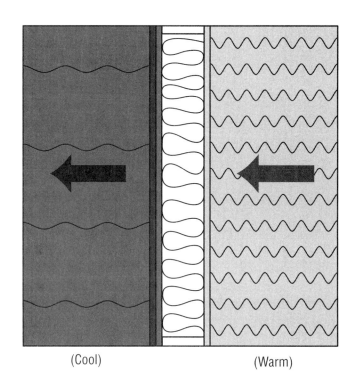

(Cool)

(Warm)

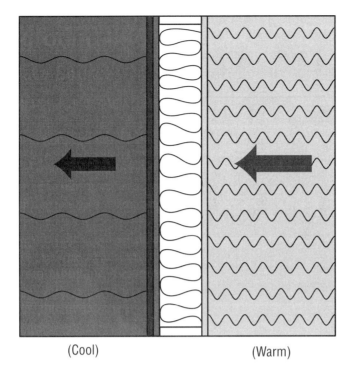

(Cool)   (Warm)

Figure 1-35 Cross-Section of a Wall with Coded Regions and Arrows The reduction in arrow size indicates the role of the wall insulation in retarding heat flow from the interior space to the exterior space.

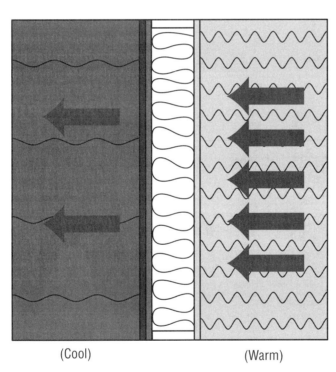

(Cool)   (Warm)

Figure 1-36 Cross-Section of a Wall with Coded Regions and Arrows Instead of a reduction in arrow size, a decrease in the number of arrows from the interior space to exterior space represents the role of the wall insulation in retarding heat flow.

flow of heat from the interior to exterior space. (We start with the assumption that it is heating season and the inside of the building is being heated.)

One question to ask of the previous illustration is: How is the role of the insulation being represented in this building system? Reviewing what was stated previously, insulation retards the flow of heat (energy) from the hot to cold region. Do the arrows indicate any slowing of this movement? The arrow is the same size. One alternative would be to shrink the arrow on the exterior side of the wall (Figure 1-35).

Does the arrow need to change color? If the idea is to represent the flow of heat, then the arrows on both sides of the wall could stay the same color (Figure 1-36). Also, if the goal is to show that insulation slows the movement of the amount of heat to the outside, then maybe total number of arrows rather than the size of the arrow might be more appropriate. It could be that a smaller arrow may be thought to represent slower movement. Is that an appropriate way of thinking of the interaction of heat energy and insulation?

Another issue may be with the use of arrows. Insulation primarily retards the rate of conduction, which only indirectly involves the movement of air. "Drafts" in a building are actual leaks in the **building envelope** (e.g., an open window), whereas conduction heat loss happens through a sealed wall. An alternative would be to represent the heat quantity as spheres (Figure 1-37).

Both the spheres and the arrows in the two previous figures show the amount of heat energy on either side of the wall, but not what is happening within the wall itself. One way to do this is to return to the use of arrows, but show the arrows penetrating the wall (Figure 1-38). Using this method, you can show that the

Figure 1-37 Cross-Section of a Wall with Coded Regions and Spheres Spheres, instead of arrows, are an alternative way to represent heat flow.

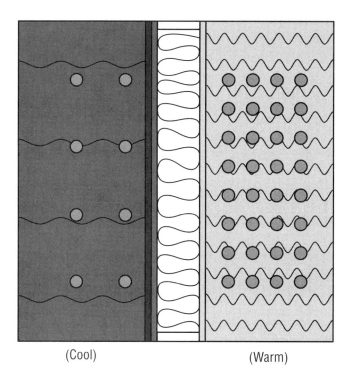

(Cool)   (Warm)

Figure 1-38 Cross-Section of a Wall with Coded Regions and Penetrating Arrows The incorporation of many smaller arrows that penetrate the wall to differing depths indicates the effectiveness of the insulation to contain the interior heat.

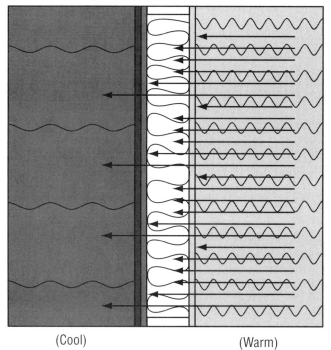

(Cool)   (Warm)

sheathing is retarding some of the heat movement, but the fiberglass insulation is doing most of this work. Going back to arrows, of course, brings up the issue of misrepresentation of conduction.

Another option is to combine more than one of the representational techniques previously shown. In Figure 1-39, the thinner arrows penetrating the wall are combined with the spheres. An advantage of this approach is using the arrows to indicate the dynamic movement of heat energy out of the building and the spheres to show the relative balance of heat energy on the two sides of the building wall.

One representation we haven't explored is what happens when there is not any fiberglass insulation in the wall. A figure like this can be used as a comparison to walls with insulation. Such pairing of images is a powerful tool.

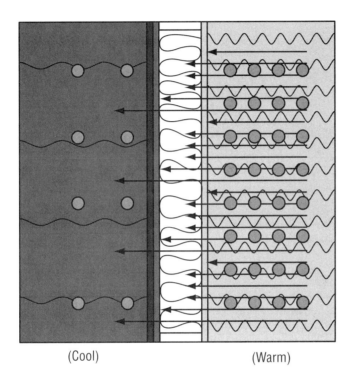

(Cool)   (Warm)

Figure 1-39 Cross-Section of a Wall with Coded Regions, Penetrating Arrows, and Spheres A combination of arrows and spheres to represent heat flow through the wall is an effective way to illustrate the scientific concept of conduction.

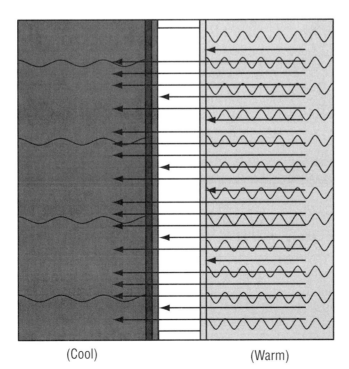

(Cool)   (Warm)

Figure 1-40 Cross-Section of a Wall without Insulation Without insulation, the wall allows more heat conduction, as illustrated by more arrows that penetrate the wall completely when compared with Figure 1-38.

A logical extension of this unit is to move beyond static graphics to representing this insulation system using dynamic graphics. Dynamic graphics provide a whole new set of opportunities for solving this communication problem. You need to think about:

- Which of the symbols are still appropriate in a dynamic graphic?
- Which symbols will stay static and which will change?
- When will the symbols change and how?

Whether it is static or dynamic graphics, the goal is to communicate information about your system to the intended audience. You decide which tools are available and are appropriate for the task. You then decide how to design your graphics to meet the goals of the problem.

Key Terms

insulation	composite board	building envelope
fiberglass	drywall	

III. Student Materials

Use Project 4 Design Brief: Communicating Insulation Technology Principles to complete the project.

UNIT 1: PROJECT 4   DESIGN BRIEF: COMMUNICATING INSULATION TECHNOLOGY PRINCIPLES

In this problem, you are going to be communicating basic principles of insulation and insulation technology to decision makers within the biotechnology company. Your goal is to help them make decisions about what kind of insulation technology should be used and assist in the calculations of the cost trade-offs of more insulation versus ongoing heating costs. You are representing a construction company submitting a proposal to the biotechnology company for constructing their new facility.

To design the graphics, the key concepts to be communicated need to be identified.

- The purpose of insulation is to retard the movement of heat (heat transfer). This transfer of heat is normally thought of as the movement from a high temperature object or area to a lower temperature object or area. The concept of heat and the movement of heat (energy) from one location to another needs to be represented.

- The different materials used in construction of the building that play a role in insulating the building need to be represented, as do any other structural materials that house/hold the insulation.

- One critical quality of insulation is how much heat passes through it in a unit of time. Does time have to be represented and if so, how?

- There are three main types of heat transfer: conduction, convection, and radiation. Conduction is the main cause of heat loss through the building walls and ceiling and needs to be represented in this graphic communication.

Part 1: Wall Section

Let's start with representing the wall of the building. The building envelope consists of the barrier between the inside and the outside and typically consists of the walls, ceiling/roof, and ground floor/foundation of the building. For the purpose of this project, we will primarily be concerned with the walls, though the ceiling/roof and ground floor/foundation may very well use a different kind of insulation technology. A wall will typically consist of a layer on the outside of the building, a layer on the inside, and the space in-between. The space in between can hold systems such as plumbing or electrical, but more importantly for this discussion, it also typically holds insulating material. The outer layer and inner layer need to be represented differently because they are made of different materials that serve different purposes (e.g., the outer layer has to be weatherproof while the inner layer does not).

Explore at least four different graphic symbols for each of these elements of the wall. In designing these symbols, consider:

- Are there any existing standard symbols used by the architectural/construction industry for these materials?

- How will they look when they are put together in a graphic? Is there enough contrast to distinguish between them? Is color needed, or can they be black and white?

- Will text labeling be needed to identify the different parts?

- What view is needed to show these different materials? A cross-section? A pictorial view of the wall?

Part 2: Representing Heat Flow

To represent heat flow, you need to identify areas of high heat and low heat (cold). In addition, you have to represent the flow of heat energy from the area of high heat to the area of low heat. Beginning with the wall section you created in Part 1, continue work on the graphic to represent the following:

- Areas of high and low heat. Can color be used to help represent this?

- Heat energy. How should this be represented? As a unit?

- Flow of heat from the high heat area to the low heat area. How do you represent movement in a static graphic?

Put these elements on different layers in the graphic so that the wall section can be alternately displayed with and without these additional graphic elements.

Part 3: Representing the Insulating Value of Different Materials

Different building materials retard heat flow (insulate) to different degrees. This is referred to as a material's R-value. For example, the sole purpose of some building material is to insulate (e.g., fiberglass batting). Other structural building materials such as wall studs also provide some insulating value. On new layers in the graphic, depict how different materials restrict the flow of heat to differing degrees.

PROJECT 5
Animating Insulation Principles

I. Project Lesson Plan

1. Project Description

You will create a 2D animation of how insulation retards the flow of heat energy. Starting with the 2D static graphics created as part of the introductory level projects, you will represent the changes in the insulation system over time by using 2D animation techniques.

2. Learning Objectives

- You will understand how animations are created by making small, incremental changes to graphic elements over a series of images, or frames.
- You will create objects to animate, both by drawing them in the animation package and importing them from other graphics packages.
- You will create an animation by translating objects over time.

3. What You Need

- Design brief
- 2D animation software (e.g., Flash, Swish, RAVE)

Optional

- PowerPoint presentation software
- Printer
- 3D modeling and animation software (e.g., 3D Studio Max)

II. Background Information

1. Planning Your Visualization: Principles of Insulation

In this problem, you are going to be communicating basic principles of **insulation** and insulation technology to decision makers within the biotechnology company to help them make decisions about what kind of insulation technology should be used and assist in the calculations of the cost trade-offs of more insulation versus ongoing heating costs. For example, you may represent a construction company submitting a proposal to the biotechnology company for insulating their new facility.

To design the graphics, the key concepts to be communicated need to be identified:

- The purpose of insulation is to retard the movement of heat (heat transfer). This transfer of heat is normally thought of as the movement from a high temperature object or area to a lower temperature object or area.

Heat is directly related to the internal energy of both objects/areas involved, according to the First Law of Thermodynamics. For the purpose of this communication, a detailed knowledge of the First Law of Thermodynamics is not needed. However, a knowledge of the concepts of heat and the movement of heat (energy) from one location to another is necessary.

- The different materials used in construction of the building that play a role in insulating the building need to be represented, as do any other structural materials that house/hold the insulation.

- One critical quality of insulation is how much heat passes through it in a unit of time. How will an animation help represent the dimension of time?

- There are three main types of heat transfer: **conduction, convection,** and **radiation**. Conduction is the main cause of heat loss through the building walls and ceiling. How will conduction be represented?

The **building envelope** consists of the barrier between the inside and the outside, and it typically consists of the walls, ceiling/roof, and ground floor/ foundation of the building. For the purpose of this project, we will primarily be concerned with the walls, though the ceiling/roof and ground floor/foundation may very well use a different kind of insulation technology. A wall typically consists of a layer on the outside of the building, a layer on the inside, and the space in between. The space in between can hold systems such as plumbing or electrical, but more importantly for this discussion, it also typically holds insulating material. The outer layer and inner layer need to be represented differently because they are made of different materials that serve different purposes (e.g., the outer layer has to be weatherproof while the inner layer does not).

How would we represent this wall? Each element will need to be represented with its own symbol. Symbols need to be differentiated from each other by changes in visual characteristics such as color or shape. You can also take advantage of symbols and colors that target audiences use in their daily situations.

Insulation primarily retards the rate of conduction, which only indirectly involves the movement of air. "Drafts" in a building are actual leaks in the building envelope (e.g., an open window), whereas conduction heat loss happens through a sealed wall. As mentioned before, we will primarily be concerned with conduction of heat through the wall. Conduction happens when molecules of two different materials collide, and energy, is transferred between the materials. Warmer materials have more thermal (kinetic) energy and transfer some of this energy to the colder material. Different materials have different conductivity properties—how resistant they are to accepting this thermal energy and passing it on. In the case of our wall, we've purposely used material that has high resistance to thermal conductivity, insulation, to slow the movement of thermal energy. The wall, in fact, is made up of multiple materials of differing levels of thermal resistance. Metal studs and single pane glass have low resistance, whereas wood studs and drywall have higher resistance. Insulating materials such as fiberglass batting and solid foam sheeting have even higher thermal resistance. You will need to represent the differences in thermal conductivity between these materials.

For additional examples of how these principles might be represented, see the background material in Project 4.

2. Animation Principles

The primary difference between the static graphics you created in the earlier projects for this unit and the dynamic graphics you are going to create is that **animations** change over time. You are using an animation rather than a static graphic because some element in your visualization is best represented by change over time. What changes in your visualization and how it changes are key elements in your design. Elements that might change include blocks of text; the background; geometric shapes, such as circles; and more complicated forms such as representations of humans, animals, or machines. These elements are often called *actors*, or characters in your animation. The collection of elements plus the background is called the *scene*.

Typical changes to an element to represent in an animation include:

- Movement (translation)
- Rotation
- Size (scale)
- Shape (non-uniform size change)
- Color
- Transparency (fading)
- Appearing/disappearing

With all types of animation, you have to decide how the change will take place over time. The animation is composed of a series of frames (usually hundreds or thousands) with different elements changing relative to the frame before or after it. Using a **timeline**, you indicate what each element in the scene is doing at different key points in time. The time will represent a certain frame in the animation, depending on the number of frames shown per second. This may mean indicating where elements are located, how they are oriented, what color they are, or whether they are even to appear at this time.

Depending on the software you are using, you may also have different tools or techniques you can use to create these changes that take place between the frames. The simplest form of animation is called **cell-based** animation. This technique dates to the earliest forms of animation. Using this technique, you would draw or paint each cell (or frame), changing one or more elements only slightly between each frame. With between 10 to 30 frames being shown every second in an animation, the changes between each frame are slight. The more frames shown per second and the smaller the changes between the frames, the smoother the animation.

At 30 frames per second, a 1-minute animation would have 1800 frames. You could imagine how long it would take to draw each individual frame, and most animation tools use techniques other than cell-based animation. First and foremost, animation packages are usually **object-oriented**. This means that each element, or character, can be considered either a single object or a **hierarchy** of objects. You can then control properties of the object, such as its location, size, or color. An example hierarchical object might be a person. With a hierarchical object, you can move the whole person, or rotate their arms or legs, or have arms and legs rotate while also moving with the rest of the body.

Another important tool animation packages have is the ability to define **keyframes**. With keyframe-based animation, you specify how an element is to look at one point in a timeline and how it is to look at a later point (many frames later) while the software determines all of its in-between states. So, for example, if a ball is to move 10 cm across the scene over 20 frames, then the animator would define keyframes at frame 1 and frame 20, while the software would automatically move the ball 0.5 cm in each of the intermediate frames (2 through 19). Because this technique automatically determines the in-between states, it is often also called *tweening*. Keyframes can not only define changes in location, but also shape, color, or other properties of the object.

Tweening of motion will usually move the object along a linear path. What if you want the object to move along a complex curved path? In that case, you would want to use **path-based** animation tools. With this technique, you would define a path, usually by drawing a spline curve, and then attach an object to it, indicating the beginning and endpoint. With both keyframe and path-based animation tools, you can also vary the rate at which the object changes between the beginning and end state. Often you will want to either accelerate or retard the rate of change near the beginning and end. This control is often called *ease-out* (of the beginning state) and *ease-in* (to the ending state).

3. Storyboarding

Successful communication of how thermal energy is transferred through a wall with an animation will be determined in large part by the quality of your communication design. Even more than with static graphics, a successful animation requires careful planning prior to using the software to create the animation. Figure 1-41 shows an example of what a few frames of a storyboard might look like for an insulation animation. Note that the frames in this storyboard contain both actual text and graphic elements that will be seen in the animation and notations/instructions as to how to create the animation. If possible, it can be helpful to place the text notes outside of the storyboard frame so that they don't clutter the frame or set confused for text elements that will appear in animation. Other action notations, such as arrows showing movement, will probably need to be shown in the frame but should be drawn in such a way as to indicate to you and others that it is not to be included in the animation.

In creating your storyboard, make sure that you include the following features:

- A drawn or written representation of all important elements that are to appear in this portion of the animation. These elements are often given a name to refer to. These names can also be used to name the objects in the animation package.

- Notations for the color and other rendering information for all new elements that have not appeared in a previous frame of the storyboard.

- Notations indicating the transformation (e.g., movement, fading) that elements are to undergo in this part of the animation.

- Notations for any other key change (e.g., background change, audio track changes).

You can speed the drawing of repetitive elements in a storyboard by drawing on paper thin enough that you put a finished frame underneath your new frame and trace the common elements. Similarly, if you have elements that don't change for multiple frames, you can photocopy or scan and print these elements on multiple frames and then finish drawing and notating the new elements.

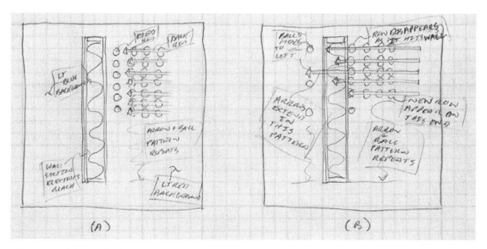

Figure 1-41 Example Storyboard Frames These frames depict the key elements in the animation and notations about their properties and how they change between key frames.

(Courtesy of the author)

Storyboards are a great problem-solving tool. Your animation is telling a story, and just as with writing a story, you have to make sure that all the pieces are logically organized, and that you've introduced all of the important ideas and presented them in a way that communicates them effectively. By representing graphically what you are planning to create, you can make sure you have not forgotten any key elements and how these elements are going to transform from one part of the animation to another. How many separate frames should a storyboard have? It really depends on how long your animation is, how many actors/elements you have in your animation, and how much these elements change over the course of the animation.

Storyboards are also a critical tool when more than one person is working on the animation. The storyboards provide a common, graphic document of what is going to happen at what point in the animation. It greatly reduces confusion and misunderstandings and helps the team to delegate responsibilities for the project work.

A complex animation may require a **flowchart** to show an even higher level representation of the animation, rather than show the details of key frames in the animation. It indicates in both words and pictures what the key elements are and what they are doing in each phase of the animation. A flowchart is particularly important if you create an interactive piece where viewers can access multiple animations and screen designs in random order.

Key Terms

insulation	animation	hierarchy
conduction	timeline	keyframe
convection	cell-based	path-based
radiation	object-oriented	flowchart
building envelope		

III. Student Materials

Use Project 5 Design Brief: Animating Insulation Principles to complete the project.

UNIT 1: PROJECT 5   DESIGN BRIEF: ANIMATING INSULATION PRINCIPLES

In this problem, you are going to be communicating basic principles of insulation and insulation technology to decision makers within the biotechnology company to help them make decisions about what kind of insulation technology should be used and assist in the calculations of the cost trade-offs of more insulation versus ongoing heating costs. You are representing a construction company submitting a proposal to the biotechnology company for constructing their new facility.

Unlike your earlier presentation, this presentation will be an animation at least 20 seconds in length. To design the animation, the key concepts to be communicated need to be identified:

- The purpose of insulation is to retard the movement of heat (heat transfer). This transfer of heat is normally thought of as the movement from a high temperature object or area to a lower temperature object or area. The concepts of heat and the movement of heat (energy) from one location to another need to be represented.

- The different materials used in construction of the building that play a role in insulating the building need to be represented, as do any other structural materials that house/hold the insulation.

- One critical quality of insulation is how much heat passes through it in a unit of time. Does time have to be represented and if so, how?

- There are three main types of heat transfer: conduction, convection, and radiation. Conduction is the main cause of heat loss through the building walls and ceiling and needs to be represented in this graphic communication.

PROJECT 6
Multimedia Presentation of Insulation Properties

I. Project Lesson Plan

1. Project Description

You will build on the animation created in Project 5 by demonstrating the heating costs related to different levels of insulation. These animations will explore alternate ways of representing the flow of heat across the insulation barrier. In addition, these animations will be merged with graphs showing how insulation values and the difference in inside and outside temperatures translate into cost to heat a space.

2. Learning Objectives

- You will use an existing animation as a starting point, create new animations that show insulation principles using techniques of movement, changing size, and changing color.
- You will modify the animation by having text fade in and out to notate important elements in the animation and to pause at key points.
- You will create graphs, export them as bitmaps, and integrate them into the animation.

3. What You Need

- Design brief
- 2D animation software (e.g., Flash or Swish)

Optional

- PowerPoint presentation software
- Printer
- 3D modeling and animation software (e.g., 3D Studio Max)

II. Background Information

1. Heat Loss Calculations

The background material in Project 5 introduced the basic concepts of how the process of **convection** moves heat energy from the warm region to the cold region and how **insulation** works to retard this process. If you own a home or run a business, you will also be interested in exactly how much heat energy is likely to move from inside your building (the warm area in winter) to the outside because you are paying for the cost of producing this heat energy.

As outlined previously, heat loss is an ongoing active process, so the best way to think about it is as the amount of heat energy that passes through a barrier (the wall) per unit time. There are many ways of representing a unit of heat energy. A popular unit of measurement in the United States is **British thermal units**, or Btus. Therefore, the amount of heat loss over time can be represented as Btu/hour. Because the loss of heat is across a barrier, you also need to know how big your barrier is. In this case, you want to know the surface area of all of your walls and ceilings that are exposed to the outside air temperature. We will ignore the floor as a source of heat loss, even though some heat is being lost through it. The easiest way to get a rough calculation of the surface area of a single room is to multiply the width and height of each wall exposed to the outside, and multiply the two dimensions of the ceiling, if it adjoins an unheated attic or roof. So if a room is 20 feet by 10 feet with 9 foot high walls, the total surface area of interest is:

Dimensions (ft)	Surface Area (sq ft)
20 x 9 =	180
20 x 9 =	180
10 x 9 =	90
10 x 9 =	90
20 x 10 =	200
Total	**740 sq ft**

Now, to calculate the heat lost from the room, you will need to know how quickly heat is being lost from the room. In theory, if there were no barrier at all, then the inside and outside temperatures would equalize quickly. In fact, you have walls with insulation in them that provide some level of resistance. The **R-factor** of the insulation provides a measure of this resistance. This R-factor uses a rather complicated unit of measurement:

$$(\text{sq ft}) \times (\text{deg F}) / (\text{Btu/hr})$$

The higher the R-factor, the lower the amount of heat energy that will pass through the barrier in an hour.

The final factor we have to consider is the difference between the inside and outside temperatures. The bigger the difference between the two temperatures, the more "pressure" there is for the heat energy to pass from the hot to cold region. Let's take all of the components we've discussed and put them in a formula to measure the heat loss for our room. We'll use the room we calculated earlier, a difference of outside to inside temperature of 15 degrees, and assume we have an R-rating of 19 in the wall:

Heat loss rate = (surface area) × (temperature difference) / R-rating

Heat loss rate = (740 sq ft) × (15 deg F) / (19 (sq ft) × (deg F) / (Btu/hr))

Heat loss rate = 584 Btu/hr

To summarize, the:

- Larger the surface area, the faster you lose heat
- Bigger the temperature difference, the faster you lose heat
- Higher the R-factor, the slower you lose heat

This calculation will tell you how fast you will loose heat in an hour for a given temperature difference, but it doesn't go far enough to really help you understand the total amount of energy you are likely to lose over a heating season. To do this, you need to know the average **degree days** for your location. This degree-day value was calculated in Project 3 and represents a summation of the average temperature difference per day for the entire heating season. The calculation might go like this:

Date	Inside	Outside	Difference
Jan 1	72	12	60 deg F
Jan 2	72	20	52 deg F
Jan 3	72	22	50 deg F
. . .	. . .	. . .	. . .
Apr 30	72	68	4 deg F

You would now add up all of these differences to come up with the total degree days. What date you set and end is, of course, somewhat arbitrary and represents an estimate. An even rougher approximation was made in Project 3 by using monthly, rather than daily, averages.

Now, if you are going to use degree days to do your calculations, you will want to know how much heat is being lost for a single degree of difference over a day:

Heat loss rate = [(740 sq ft) × (1 deg F) / (19 (sq ft) × (deg F) / (Btu/hr))] × 24 hr/day

Heat loss rate = 935 Btu/degree day

You can now multiply this heat loss rate for a single degree by the total number of degree days for your location. Going back to Project 3, Chicago had 8301 degree days with an inside temperature of 72 degrees Fahrenheit. Your energy loss for the year would then be:

Heat loss per year = (935 Btu/degree day) × (8301 degree days)

Heat loss per year = 7,759,250 Btu or 7.8 million Btu

All right, now the big question is how much those 7.8 million Btus are costing you. If you are heating with natural gas, you might be paying $10/million Btu:

Yearly cost = (million Btu needed) × (heating cost/million Btu)

Yearly cost = (7.8 million Btu) × ($10/million Btu) = $78

So, for this one room, you would be paying $78/year to heat it. Sounds like a lot? Actually, it would be higher. This calculation assumes that 100 percent of the natural gas gets converted into useful heat. More realistically, the efficiency is probably closer to 70 percent. So, a more accurate formula would be:

Yearly cost = (million Btu needed) × (heating cost/million Btu) / (heating efficiency)

Yearly cost = (7.8 million Btu) × ($10/million Btu) / .70 = $111

2. Enhancing Animations

The design brief for the current project asks for a number of enhancements. All of the enhancements build off of the initial animation design by incorporating more sophisticated visualization techniques.

Text can be a powerful tool for enhancing a technical animation. While the main focus of these projects is to use graphics to communicate technical and scientific information, judicious use of text can help communicate and reinforce ideas being presented graphically. It is particularly powerful as a labeling tool, where a few choice words can name components or concepts. Text can be included in animations as another element, or actor, in the animation. A key is picking the appropriate point in time and location in the scene for the text to appear and, then, when to remove it. Text, as another element, needs to be seen, but you also don't want it to be a distraction. This means you will need to design its size, font, and color so that it is visible. It also means that you will want to plan its time visible in the scene during a time that does not compete with other important dynamic activity in the scene or at a location that covers up important elements.

Often, you will want to extend the activity of the elements that the text is referring to so that there is time to both read the text and watch the graphic elements it is referring to. You will often need to try out a number of different combinations of time length and entry/exit points for the graphic and textual elements. In some cases, the graphic element will freeze while the text shows; other times, loop its dynamic activity. Another way to handle text is to have it appear on the outer edge of the animation and hold for the entire length of the animation. This way the text can be referred to at any time.

Text and other static graphic elements are often combined with the animation to create a larger scale **multimedia** presentation. This presentation might contain a series of shorter animations, each paired with static text and graphic elements that support the animation. You can use a software tool such as PowerPoint to create a series of slides, each of which contains multiple static and dynamic elements. By using slides, the viewer can control the pace at which each slide is shown, keeping it visible long enough to fully inspect the static elements and to replay the animation as many times as necessary. Another option would be to use an animation tool that allows for dynamic, interactive control. Multimedia packages such as Flash allow you to create interactive buttons that allow the viewer to control which static and dynamic elements appear and for how long.

Key Terms

convection	British thermal units (Btu)	degree days
insulation	R-factor	multimedia

III. Student Materials

Use Project 6 Design Brief: Multimedia Presentation of Insulation Properties to complete the project.

UNIT 1: PROJECT 6   DESIGN BRIEF: MULTIMEDIA PRESENTATION OF INSULATION PROPERTIES

In this problem, you are going to be communicating the principles of insulation and insulation technology to decision makers within the biotechnology company to help them make decisions about what kind of insulation technology should be used and assist in the calculations of the cost trade-offs of more insulation versus ongoing heating costs. You are representing a construction company submitting a proposal to the biotechnology company for constructing their new facility.

Unlike your earlier presentations, this multimedia presentation will be a series of animations, each at least 10 seconds in length. These animations will be coupled with static text and graphics to communicate the costs of not adequately insulating a manufacturing building. To design the presentation, the key concepts to be communicated need to be identified.

- The purpose of insulation is to retard the movement of heat (heat transfer). This transfer of heat is normally thought of as the movement from a high temperature object or area to a lower temperature object or area. The concept of heat and the movement of heat (energy) from one location to another needs to be represented.

- The different materials used in the construction of the building that play a role in insulating the building need to be represented, as do any other structural materials that house/hold the insulation.

One central interest in this animation is those factors that affect the cost of heating the manufacturing space. These include the:

- Amount of insulation (represented by the R-factor).

- Amount of surface area being insulated.

- Difference between indoor and outdoor temperatures (and, therefore, the number of degree days).

- Cost of heating fuel and the efficiency of the heating system.

- Combination of animations, static graphics (including charts and graphs), and text to convey these relationships.

Your multimedia presentation should first be storyboarded and then reviewed and approved by your teacher.

UNIT 2
Communications Technology: Introduction to 3D Modeling

Unit Overview

I. Introduction

While many scientific and technical concepts can be expressed as two-dimensional images, others are better expressed in three dimensions. In particular, when you need to represent complex three-dimensional (3D) geometry, 3D modeling tools greatly enhance your technique. While 3D objects can be represented as a pictorial projection using 2D graphics tools, this process is inherently slow and difficult. When you then want to animate the 3D object changing in shape or location, creating multiple pictorial views becomes very difficult. For this reason, 3D modeling tools are the tools of choice to represent static 3D objects and, specifically, animated 3D objects.

In this unit, pump technology—those types of pumps used in medical applications—are going to be the focus of our investigation. Through this activity, you will investigate how the specific needs of different medical applications drive the selection of different pump

technologies. Particularly, you will examine the mechanisms and their underlying geometry and materials and evaluate how particular designs meet the various medical applications.

Imagine you need to represent the commonalities and differences between the types of pumps used in medical applications. While 2D graphics can be used to represent these pumps, it would be difficult to fully represent the geometries underlying the key technological components or the arrangements of component parts of the entire pump assemblies. Animating the functions of these pumps would further add to the difficulties for 2D representation. For these reasons, 3D modeling tools would be optimal for the representation of the geometry underlying medical pump technology.

II. Unit Learning Goals

- To develop an awareness of the medical applications of pumping and hydraulic technologies
- To get an overview of 3D computer modeling and animation tools and techniques
- To use 3D object-oriented graphics software to represent different types of pump technologies

III. Projects

Introductory Projects

PROJECT 1: Pump Technology

In this project, you will explore the difference between basic types of pumps used in medical devices. In particular, you will study piston (syringe) pumps, diaphragm pumps, and peristaltic pumps that use both rotary and linear motion. You will also begin planning the model parts that will be constructed in Project 2.

PROJECT 2: Part Modeling

In this project, you will learn the basics of 3D solid modeling techniques by using the basic modeling commands to create the individual component parts of a syringe pump. Because in Project 3, these parts will be brought together into a complete assembly of the syringe pump, initial planning is necessary to make sure the parts will fit together.

PROJECT 3: Assembly Modeling

In this project, you will take the component parts built in Project 2 and create a complete assembly of the syringe pump. This project will allow you to test the planning and design carried out in Project 2 and determine whether the parts fit together properly. Project 4 will allow you to test the functionality of the assembly by animating it.

PROJECT 4: Animation

In this project, you will take the assembly created in Project 3 and animate it. The animation should depict the functionality of the syringe pump and also provide an opportunity to test whether designed part design geometry allows for the proper movement and functioning of the parts.

Intermediate Projects

PROJECT 5: Rendering and Deformation

In this project, you will explore rendering techniques to help visualize pump functionality. Specifically, you will look at how surface properties (e.g., color and reflectivity) and lighting are used to help communicate the pump's geometry and functionality. In addition to rendering, you will also study how geometry deformation plays a role in some pump technologies. You will learn about the relationship of general pump technologies and how deformable materials are used in different types of pumps. In this project, you will use modeling/animation deformation techniques to demonstrate these designs.

PROJECT 6: Camera Viewpoints and Particle Systems

In this project, you will explore advanced modeling and animation techniques as a way of visualizing medical pump functionality. You can use alternate camera viewpoints to highlight particular parts of the pump or take the viewpoint of a blood cell going through the pump. Particle systems are an advanced modeling technique used to simulate the complex movement of fluids and gasses and those elements that would move through them. For medical pumps, this might be the flow of blood cells through the pumps.

Advanced Project

You will complete an independent project using visualization tools. You will research a new topic about medical pumps and expand the modeling and animation techniques covered in this unit. The objective of the advanced level is for you to further your skills in integrating research, problem solving through the design brief approach, and presentation. Your teacher will work with you to negotiate the topic, time allocated to the project, and design constraints.

IV. Unit Resources

The Resource index at the end of this book contains a listing of all resources associated with the unit. Included are relevant Web sites, books, and other publications. The Glossary provides definitions for all key terms listed in each project.

PROJECT 1
Pump Technology

I. Project Lesson Plan

1. Project Description

In this project, you will explore the differences between basic types of pumps used in medical devices. In particular, you will study piston (syringe) pumps, diaphragm pumps, and peristaltic pumps that use both rotary and linear motion. You will also begin planning the model parts that will be constructed in Project 2.

2. Learning Objectives

- You will define different types of pump technology.
- You will gain a historical perspective of pump technologies.
- You will gain an appreciation of how pumps are used in medical applications.
- You will synthesize the information collected through research by creating sketches of 3D model parts.

II. Background Information

1. Pump Technology

The primary purpose of **pumps** is to move liquid or gas from one space to another. In special cases, the pump may also move a *slurry* (solids suspended in a liquid or gas). This movement often happens through enclosed channels such as pipes or ductwork. Pumps typically use some type of motor to power a mechanism that moves the gas or liquid. The mechanism may use rotary or linear movement or a combination of both. Usually, there will be either valves or another mechanism that keeps the fluid from moving in more than one direction through the pumping system.

Pumps can work by either negative or positive pressure when moving fluids. Vacuum pumps are often used to suck the contents from the stomach and lungs. Positive gas pumps are used to provide clean air or special gaseous mixtures (e.g., oxygen) into patients' lungs.

The material for this unit provides a brief overview of different types of pumps, with particular emphasis on pumps used in medical applications. In addition, design considerations unique to the medical field will be discussed.

a. Kinetic

One type of **kinetic pump**, centrifugal pumps, take fluid near the central axis or rotation and, as the impeller rotates, the fluid is forced to the outside at high speed. The kinetic energy can be converted to higher pressure by restricting the area that the fluid flows through. This device can accommodate liquids, gases, and fluids carrying small particles. **Axial** flow devices carry the fluid in the same direction as the rotating axis. An electric fan is a simple example. In a closed chamber, the pressure and flow of the fluid can be controlled.

b. Electromagnetic

Fluids that are good electrical conductors can be moved with **electromagnetic pumps** through manipulation of the electromagnetic field. Electrical wiring wrapped around a pipe creates a magnetic field force in the direction that the fluid should move. Because there are no moving parts or seals where fluid can leak, it is a good choice for this type of dangerous application. This type of pump moves sodium coolant used in nuclear power plant cooling.

c. Positive Displacement

Positive displacement pumps use positive pressure to move fluids. One of the oldest type of pump, the **screw pump**, was used to move water through irrigation systems as long ago as the third century BC. This type of pump is often referred to as the **Archimedes pump** (Figure 2-1 and Figure 2-2).

Gear and lobe pumps are commonly used to move thick, viscous liquids such as oil. A gear pump moves the liquid between the meshing teeth of two gears.

Figure 2-1 Screw Pump Woodcut This woodcut illustrates an Archimedes' Screw pumping water from a stream to a bucket. It comes from a 1522 edition of Virtruvius's *De Architectura*, by Fra Gioconda.

Figure 2-2 Screw Pump Schematic

(Used with permission, Bureau of Reclamation)

As the teeth come together, the liquid is trapped, compressed, and pushed out the other side. Similarly, a lobe pump captures the liquid between a rotating lobe and the side wall of a chamber before it is pushed out the other side (Figure 2-3).

Figure 2-3 Lobe Pump The lobe pump captures the fluid against the chamber wall and pushes it through.

d. Piston Pumps

A **piston pump** uses a piston in a cylinder to push fluids (Figure 2-4). The piston has one side that encloses a chamber where the fluid comes in and leaves while the other side is usually connected to an arm that controls the movement of the piston. If the piston is going to continue to move in and out, **valves** can be used to control when the chamber fills and empties. Most internal combustion engines use a variation on this type of pump technology. The difference is that instead of having the power applied through the piston

Figure 2-4 Piston Pump Cross-sectional diagram of a simple single-piston reciprocating pump.

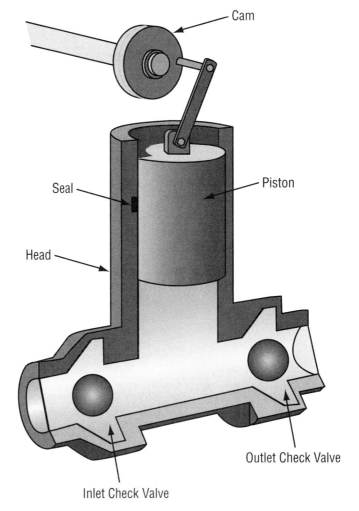

(© LC Resources Inc.)

to the gas in the chamber, with the internal combustion engine the gas in the chamber explodes, powering the piston down.

One type of piston pump used in medical applications is the **syringe pump**. Here, a syringe is designed with a plunger driven down a single time. The plunger is pushed by a shoe on a worm gear attached to gearing and a motor, or directly to a stepping motor. To provide continuous fluid flow, the pump can be set up as a double syringe unit where valves allow one syringe to fill while the other is discharging. There are also reciprocating syringe pumps that use a single chamber that alternately fills and discharges on opposite sides of the barrel. Syringe pumps are often used on volumetric infusion pumps, a common medical pump.

While syringe pumps can be accurate, they suffer from an effect called **stiction** when they are plunged at a very slow rate. Stiction is a combination of the words *sticking* and *friction*. The piston in syringes is made of a flexible material that presses against the wall of the cylindrical chamber to create a seal. This material tends to grab or stick on the wall and needs a steady push to break free and begin to slide down the wall. When pushed at a very slow rate, it doesn't ever quite break free and jerks down at an uneven rate. The other disadvantage of a simple syringe pump is that once the contents of the syringe is emptied, it has to be taken out and either refilled or replaced by hand. For continuous flow, a double or reciprocating syringe pump is needed. These types of pumps are obviously more complex and have more possible failure points.

e. Diaphragm Pumps

A variation on the piston pump is the **diaphragm pump**. Instead of having a moving piston inside of a cylindrical chamber, there is a flexible membrane (diaphragm) that is pulled and pushed with a reciprocating arm. When the diaphragm is pulled, a suction draws the fluid into the chamber and when is it pushed, the pressure forces the fluid out. As with a reciprocating piston pump, valves control the flow in and out of the chamber.

One advantage of the diaphragm pump is that there is no space between the diaphragm and the chamber where fluid can leak through as there is with a piston pump. This is an important consideration when the fluid being moved has to be kept sterile. However, the flexible material of the diaphragm will fail over time. Diaphragm pumps have been used for many centuries. **Bilge pumps**, used to pump excess water from inside boats, used diaphragm pumps with leather membranes as far back as the sixteenth century (Figure 2-5).

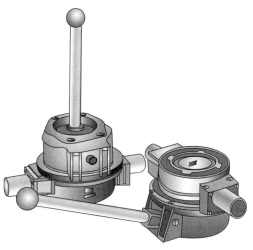

Figure 2-5 Diaphragm Pump This diaphragm pump is used in boats to manually pump bilge water.

(Courtesy Johnson Pump AB)

f. Peristaltic Pumps

Another type of positive displacement pump is a **peristaltic pump**. *Peristalsis* means "flowing contraction" and is the type of motion used by the esophagus and the intestines to move food through the digestive track. One of the most common types of peristaltic pumps uses two or three rollers that squeeze a U-shaped tube against a grooved slot (shoe) to produce the peristaltic effect (Figure 2-6). This type of pump is popular as a blood pump because of the high degree of sterility (nothing goes inside the tube) and the motion is gentle on the blood products. The qualities of the tubing used under the rollers are critical. If it is too soft, it will allow too much movement of the tube; if it is too hard, it does not allow full compression or it could strain on the roller assembly. As with the diaphragm pump, life of service of the tubing is critical. Tubing will lose its elasticity over time and/or crack and break. Lubricants such as silicone jelly are often used between the rollers and the tubing to help extend its life.

This pump has the advantage of completely isolating the fluid inside the tubing and continuous flow of fluids. However, it may not provide the same level of accuracy as a syringe pump would have pumping fluid. Often a roller pump has a secondary flow monitoring device, such as an optical eye that counts drops coming out of the tube.

A **finger pump** is another type of peristaltic pump. It uses a **cam** to press a series of fingers onto a bladder or tube. This can be done in one pass on a disposable bladder or as continuous motion on a tube. This type of pump is often used in personal portable insulin pumps.

Figure 2-6 Peristaltic Pumps The pump on the left is a two-roller pump, while the one on the right is a three-roller model.

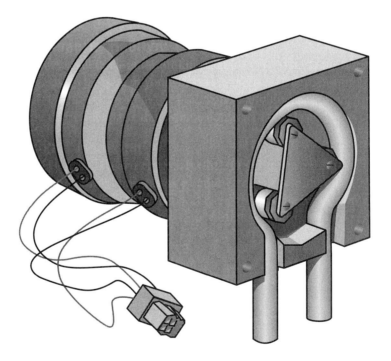

Key Terms

pump	Archimedes pump	diaphragm pump
kinetic pump	gear pump	bilge pump
axial	lobe pump	peristaltic pump
electromagnetic pump	piston pump	finger pump
positive displacement pump	valve	cam
screw pump	syringe pump	
	stiction	

III. Student Materials

Use Project 1 Design Brief: Exploring Different Medical Pump Types to complete the project.

UNIT 2: PROJECT 1   DESIGN BRIEF: EXPLORING DIFFERENT MEDICAL PUMP TYPES

Your medical technology company is about to begin the design of a new line of medical pumps to be used for infusing specially treated blood products into the bloodstream. As is the case with most pumps used in medical applications, this new line of pumps has to meet the following design criteria:

- The pumping action has to be gentle and not damage the fragile blood components (e.g., red blood cells).
- The areas of the pump that the blood products come in contact with have to be kept sterile and easy to clean.
- An accurate, constant flow of this blood product has to be infused into the bloodstream over a long period of time.

Three of the more common pump technologies are:

- Syringe pumps
- Diaphragm pumps
- Peristaltic pumps

1. Research and discuss each of these pump types.
2. Next, create sketches of each of the pump types and document the strengths and weaknesses of each pump type based on the criteria listed previously.
3. Make sure that the key components of each pump are clearly labeled using color and/or text labeling.

PROJECT 2
Part Modeling

I. Project Lesson Plan

1. Project Description

In this project, you will learn the basics of 3D solid modeling techniques by using the basic modeling commands to create the individual component parts of a syringe pump. Because in Project 3 these parts will be brought together into a complete assembly of the syringe pump, initial planning is necessary to make sure the parts will fit together.

2. Learning Objectives

- You will learn the fundamental principles of 3D modeling.
- You will be able to define the basic part modeling commands.
- You will learn how to apply basic modeling operations to create different types of geometry.
- You will gain an appreciation of how part modeling can be used in the design and visualization process.

II. Background Information

1. Geometry Creation

The beginning point of any 3D model is the creation of geometry. While modelers all differ in the approach they take to the creation of geometry, there are some general approaches that are common to all modelers. Some of the most common geometry creation processes include primitive instancing and generalized sweeps. With these methods as starting points, geometry can be modified with Boolean operations and with other more specialized operations that act on existing geometry. Finally, multiple individual geometric forms, or **parts**, can be brought together using assembly linking and building techniques.

All geometry will be created in a **virtual world**, often called a **scene**. The location of geometry in the scene is usually specified with a **3D Cartesian coordinate system**. The intersection of the three mutually perpendicular axes, X, Y, and Z, defines the origin of the scene. Geometry is placed in the scene relative to this coordinate system. This coordinate system is sometimes called the **absolute coordinate system**, because it ultimately defines the complete scene. Other, relative, coordinate systems can also be defined based on other geometry in the scene. These **relative coordinate systems** are still tied back to the absolute coordinate system.

As you model, you will want to change your view of the model with which you are working. Some modelers will allow you to have multiple views of the model at the same time, while others only have a single view window. Two of the most common types of views are **orthographic** and **pictorial views** (Figure 2-7). Orthographic views are useful for giving you an undistorted view of two dimensions of your model. This allows you to accurately click on key edges (lines) and vertices (intersections) or judge the location of features relative to each other. A pictorial view shows all three dimensions of an object and provides a holistic view of the complete model. From this view, you can see how your entire model is coming together and plan future geometry edits and view locations.

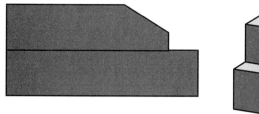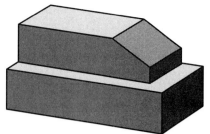

Figure 2-7 Orthographic and Pictorial Views An orthographic view (left) gives you a non-distorted view of two dimensions of your model while a pictorial view (right) will give you a holistic view of your model.

a. Primitive Instancing

Primitive instancing is an approach that allows users to specify the dimensions (parameters) of basic geometric solids. For example, if the user was creating a rectangular prism, they might specify its height, width, and depth (Figure 2-8). These specifications might be given by typing in numeric values, or interactively gesturing the dimensions in the drawing area with a mouse. Primitives typically fall into a few basic categories. **Prisms** are 2D polygons such as a rectangle, triangle, or hexagon that have been extruded into a third dimension. The results are, predictably, rectangular, triangular, and hexagonal prisms. A prism whose sides are perpendicular to the end face polygon is called a right prism. When the sides are not perpendicular, it is called an oblique prism. Circles or ellipses that are extruded are called **cylinders**. When a circle is extruded but tapers to a point, it is a **cone**. When a square or triangle comes to a point, these are square and triangular **pyramids**, respectively. When a prism, cone, or pyramid is sliced off at some point above its base such that none of the base is cut away, it is called a **frustrum**. For example, if a cone is sliced off half way between its base and its point, it is called a *frustrum* of a cone. Finally, two important special case primitives are the **sphere** and **torus**—informally referred to as a "doughnut," after the food it resembles.

b. Sweeping

A wider variety of geometric solids can be created using a method called generalized **sweeping**. With sweeping, a 2D polygon is drawn and then moved along a linear or curved path to create a 3D solid. Typically, a plane is defined in space on which to draw the 2D polygon (often called a **profile**). The plane on which the profile is drawn can be out in space or a flat surface on the model. Usually the profile has to be a closed polygon where the end of the last line segment connects with the beginning of the first segment. A closed polygon is needed to create a solid form. If the polygon is open, then the software will either close the polygon automatically or create a **surface** rather than a **solid** from the profile when it is swept out.

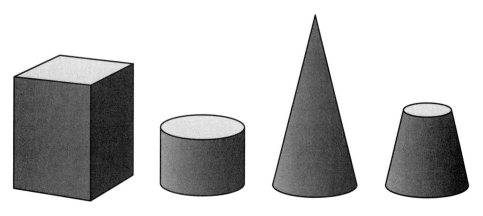

Figure 2-8 Primitives Primitives are usually classified as prisms, cylinders, cones, or pyramids. Trimming the top of any of these primitives creates a frustrum.

The profile can be swept out in a path that is either **linear** or **curved**. A linear sweep is often called an **extrusion** (Figure 2-9). Some software limits the direction of the sweep to being perpendicular to the profile. Sweeping a rectangular polygon with a perpendicular extrusion creates a right rectangular prism. Other software allows this extrusion to go at some other angle between 1 and 90 degrees. The angle could not be 0 degrees because this would not create a solid coming off the profile's plane!

Being able to create curved sweeps along paths that are not linear provides a lot of flexibility to create different solids. When defining a **circular sweep**, an axis of revolution needs to be defined (Figure 2-10). This axis will always reside in the same plane as the profile, and will either be along one side of the profile or placed some distance away. If it is placed some distance away, then it will create a hollow in the middle of the form.

Figure 2-9 Linear Sweeps A linear sweep that is at right angles to the profile creates a right prism. Linear sweeps can sometimes also be done at angles to the profile.

Figure 2-10 Circular Sweeps A circular sweep is done about an axis and can sweep a profile any number of degrees about this axis. Different forms can be created by locating the sweeping axis on or off the edge of the profile.

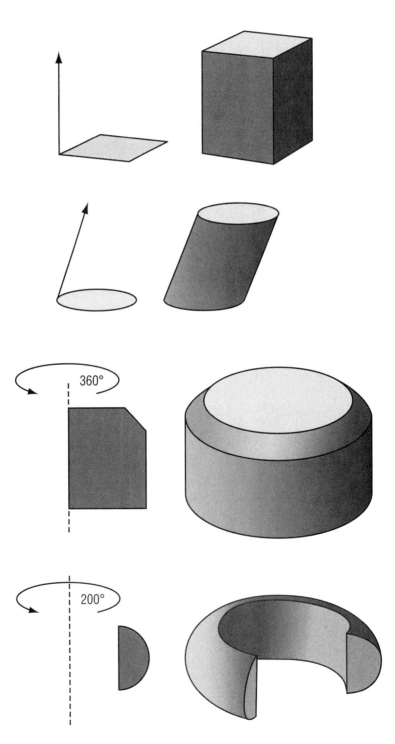

Paths other than simply linear or circular allow even more complex forms to be created (Figure 2-11). A combination of linear and curved elements or a series of curved elements can be used to define a path along which a profile is swept. This path may or may not lie in a single plane, depending on what the software supports. Finally, multiple profiles can be used to define a 3D solid. These extrusions—sometimes called **blends**—have two or more profiles that may or may not be parallel to each other. The software then decides how the swept form should transition from one profile to the next. For example, two circles of different sizes spaced apart can define a frustrum of a cone. If something other than a linear transition is wanted between the profiles, then a curved path can be drawn to describe the transition.

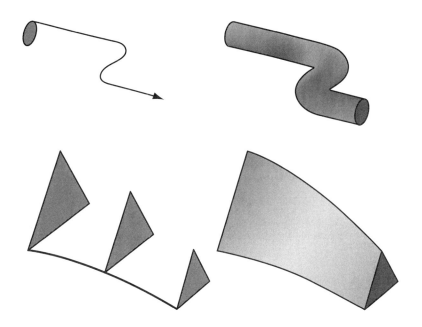

Figure 2-11 Path Sweeps and Blends Forms can also be created by sweeping a profile along a complex path. Blends merge together multiple profiles.

c. Boolean Operations

Once the initial 3D form of the model has been created, additional geometry can be added or subtracted from it in a number of ways. One of the most common approaches is to use what are called **Boolean operations** (Figure 2-12). With these operations, a separate 3D form is temporarily created that either overlaps or adjoins the existing 3D model. This new form is either added (a Boolean **Union** operation) or subtracted (a Boolean **Subtraction** operation) from the existing model. With a Union operation, any overlap between the two objects is counted only once (you don't have "double solids" where they overlap). In a Subtraction operation, the area of overlap between the forms is removed from the original model. You can also have a Boolean **Intersection** operation where the area of overlap is all that remains of the model after the operation.

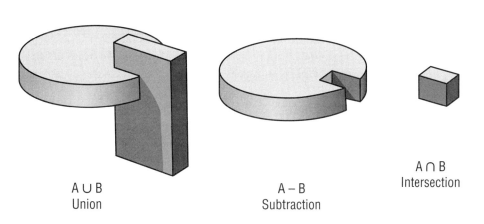

Figure 2-12 Boolean Operations Boolean operations can be used to add and subtract geometry from the model.

A ∪ B
Union

A − B
Subtraction

A ∩ B
Intersection

Depending on the modeler, the Boolean operation may cause permanent change to the original form, or you may be able to reverse the operation at a later time. Similarly, after the operation, the new form used for the operation may disappear or it may remain as its own separate form.

2. Geometry Editing

After an initial model is completed, additional modifications are usually needed to correct earlier errors or to generate design alternatives. There are many approaches to modifying the model, including the complete replacement of 3D forms you had originally created. Often, each geometry creation step or Boolean operation is saved separately. If the modeler allows, you may also be able to go back and modify the sweeping profiles or paths used to create the swept forms. Also, it may be possible to modify the location of forms used in Boolean operations. In addition to these methods, there are often tools that allow other alternatives to editing parts. These may include direct editing of the edges and vertices of the model (Figure 2-13).

Figure 2-13 Rigid Transformations Geometry can be transformed by translating, rotating or scaling it.

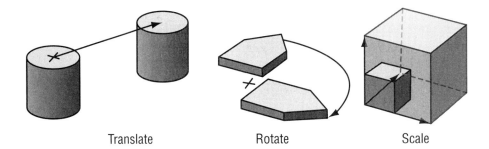

Translate   Rotate   Scale

a. Rigid Transformations

If a part is not in the proper location or orientation, it can usually be moved relative to other parts in the scene (the "virtual world" the part lives in) by applying a **rigid transformation**. These transformations are called *rigid* because they do not change the shape of the object, only its location, orientation, or overall size. Location changes are done with **translations** in the X, Y, or Z axis, whereas orientation changes are done by **rotating** the object around the same three axes. A rigid **scaling** will uniformly expand the object in the X, Y, and Z axes.

b. Non-rigid Transformations

Non-rigid transformations change, or distort, the shape of the object (Figure 2-14). One example would be a non-uniform scaling. For example, you could increase the length of a rectangular prism by scaling only in the X direction. Not surprisingly, this type of non-uniform scaling is often called **stretching**. Another type of non-rigid transformation is called **shearing**. In this case, a part of an object is translated while the rest remains in place. An example would be if a cube was sheared, causing two sides of the cube to be parallelograms. Less common is rotating a face, causing other faces to twist, and no longer be planar. Finally, more local editing can be done on an object, where a corner (vertex) or face is pushed or pulled, creating a non-rigid transformation.

c. Duplication

Another type of editing is **duplication** of parts. These duplications are usually done by applying rigid transformations to copies of parts. A linear duplication translates a copy of a part while a radial duplication rotates a part about some axis away from the part. Multiple parts to be copied at the same time constitute an **array** operation. In this case, each part is duplicated then

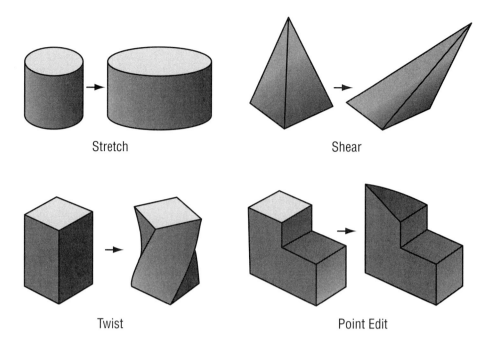

Stretch

Shear

Twist

Point Edit

Figure 2-14 Non-rigid Transformations Geometry can be transformed with non-rigid transformations in ways that distort the geometry.

transformed multiple times using the same increment. This process is repeated for as many duplicates as you want. An array using translation is called a *linear array*, while one using rotation is called a *radial array* (Figure 2-15). A linear array can be done in one dimension, creating a line of duplicated parts, while a 2D array creates a rectangular grid of duplicated parts.

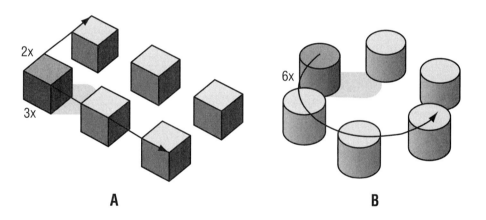

2x

3x

A

6x

B

Figure 2-15 Array Duplication A 2D linear array creates multiple copies in two dimensions (A), while a radial array creates multiple copies about an axis (B).

Key Terms

part	cone	Boolean operation
virtual world	pyramid	Union
scene	frustrum	Subtraction
3D Cartesian coordinate	sphere	Intersection
system	torus	rigid transformation
absolute coordinate	sweeping	translation
system	profile	rotate
relative coordinate	surface	scale
systems	solid	non-rigid transformation
orthographic view	linear	stretching
pictorial view	curved	shearing
primitive instancing	extrusion	duplication
prism	circular sweep	array
cylinder	blend	

III. Student Materials

Use Project 2 Design Brief: Modeling a Medical Pump to complete the project.

UNIT 2: PROJECT 2   DESIGN BRIEF: MODELING A MEDICAL PUMP

Your medical technology company is about to begin the design of a new line of medical pumps to be used for infusing specially treated blood products into the bloodstream. As is the case with most pumps used in medical applications, this new line of pumps has to meet the following design criteria:

- The pumping action has to be gentle and not damage the fragile blood components (e.g., red blood cells).
- The areas of the pump that the blood products come in contact with have to be kept sterile and easy to clean.
- An accurate, constant flow of this blood product has to be infused into the bloodstream over a long period of time.

Three of the more common pump technologies are:

- Syringe pumps
- Diaphragm pumps
- Peristaltic pumps

1. In the previous project, you researched the three types of medical pumps. Select one of the pumps for modeling using a 3D modeling system.
2. Create "process sketches," showing the steps you will take to model each part of the pump
3. Create a 3D model of each part of the pump.
4. As requested, either create printouts of one or more views of each of the parts or be prepared to show the parts on the computer.

PROJECT 3
Assembly Modeling

I. Project Lesson Plan

1. Project Description

In this project, you will take the component parts built in Project 2 and create a complete assembly of the syringe pump. This unit will allow you to test the planning and design carried out in Project 2 and test whether the parts fit together properly. Project 4 will allow you to test the functionality of your assembly by animating it.

2. Learning Objectives

- You will learn the fundamental principles of 3D modeling.
- You will be able to define the basic assembly modeling commands.
- You will learn how to apply assembly modeling operations to create different types of geometry.
- You will gain an appreciation of how assembly modeling can be used in the design and visualization process.

II. Background Information

1. Assembly Modeling

Most manufactured products are created through a process of fabricating a number of separate pieces and then assembling them into a final product, or **assembly**. This approach is used for a number of reasons. One reason is that it allows for optimizing the type of material used for each part of the product. Different parts of a product have different functions for which a particular material might be best suited. In a syringe pump, flexible plastics are used for tubing and the seal between the plunger and cylinder. A rigid clear plastic is used for the cylinder, while metal may be used for the screw drive assembly.

Assemblies are also used because doing so simplifies the manufacture of components. Could you imagine trying to shape a car body as a single piece of metal, as opposed to forming smaller subcomponents and assembling them? Assemblies also allow for easier substitution of parts to allow for alternative designs. For example, you might want to be able to substitute syringes of different sizes in the pump. Finally, assemblies allow parts to move relative to each other, meeting the functional needs of the product. For the syringe pump, the plunger needs to translate along a linear axis into the cylinder. On the other hand, the motor components driving the plunger will probably need to rotate about an axis.

Models of the virtual product are also built as assemblies for some of the same reasons that the real products are: the ease of modeling many simpler parts and bringing them together into a more complex assembly, and the ability to substitute parts to create design variations. Probably one of the most important reasons for assembly modeling is to model the actual process of assembly for the real product. By simulating the process of assembly in the virtual model, the designer can learn a lot about how the individual parts should be designed, fastened together and in what order.

2. Assembly Hierarchies

Much as individual parts are constructed as a series of modeling operations, assemblies are put together through a series of assembly operations. There are many parallels in the strategy used in modeling parts and assemblies, even if the names and actions of the commands are different. Planning your assembly before you start is just as important as planning your part modeling ahead of time. With an assembly, you will want to identify your **base part** (Figure 2-16).

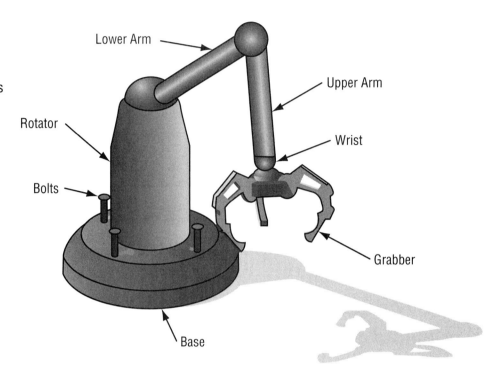

Figure 2-16 Mechanical Arm This mechanical arm is an assembly of parts, with a base part and child parts. A hierarchical representation of this assembly is seen in Figure 2-17.

While a number of parts can often be used as one, you will want to pick a part that might be considered a "core" part, around which other parts are attached. Similarly, this is often a stationary part, around and on which other parts move. Imagine building an assembly of the human body. You wouldn't start with a finger or foot as the base part. Instead, you would start with the torso so that the added parts would all move relative to the torso. Even though the foot is on the ground, logically, its movements are still controlled by the lower leg, which is attached to the upper leg, which is attached to the torso. While you could build an assembly with the foot as the base part and have the leg function logically, it would be hard to model and control the arms and head relative to the foot. The torso becomes the logical base part to assemble all of the appendages relative to it.

Assemblies are often created as **hierarchical trees**. That is, the base part is considered at the top of the tree with all other parts added to the assembly as **children** of the base part. A part in the assembly is considered a **parent** when it has parts added later in the assembly and their location and/or movement is defined relative to this part. When multiple parts are added in as children to a parent part (other than the base part), this grouping is often considered a **subassembly**. In Figure 2-17, the base part at the top of the hierarchical tree is the base of the mechanical arm. The wrist subassembly contains the wrist parent part and three child parts, the grabbers.

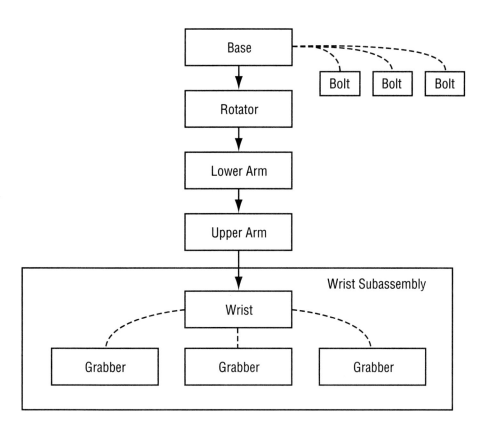

Figure 2-17 Hierarchical Representation of a Mechanical Arm A hierarchical representation has a base part and, often, subassemblies. Parts attached below another part are considered children of that parent part.

Hierarchical trees allow the modeling software to define how parts are arranged relative to each other. An assembly model will typically *point* to the individual part model geometry rather than copy and recreate the geometry in the assembly. The only new information that needs to be added is how the part is transformed (e.g., translated and rotated) into position in the assembly model. This information can be added to the hierarchical tree and will define the relationship between the child and parent parts. If the part is static, then that is the end of the transformation. If the part is meant to be a dynamic, moving part, then the assembly needs to store how the child part can be transformed relative to the parent. The next section will go into this in more detail.

Assemblies allow the same part to be brought into an assembly multiple times, avoiding the need to model the same part over and over again. Standard hardware, such as nuts and bolts, is typically used many times over in an assembly. All of the repeated parts point back to a single part model.

3. Linking with Assemblies

When parts come together into an assembly, the relation of the parts needs to be defined. As mentioned earlier, the first part brought in is the base part. Because there are no other parts, it is located and oriented relative to the coordinate system of the modeler. The next part brought in will be a child of the base part (the parent) and located and oriented relative to the base part. If the part is to be **fixed**, then it is located and oriented once. **Movable** parts are meant to move relative to their parent part.

The connection of movable parts to their parents is usually called a **joint**. Joints are defined by the types of translations and rotations the part is allowed to make relative to the parent part. Movements (transformations) are either **translations** or **rotations** along or about an axis. Since you are modeling in a 3D world, there are three possible translations—X, Y, and Z—and three possible rotations about the same axes. The number of possible translations and rotations a movable part can make is called the number of **degrees of freedom**. Figure 2-18 shows two of the most common joint movements. Notice that a hinge joint has one degree of freedom: rotation about an axis. A ball and socket joint has three degrees of freedom: rotation about three axes.

When multiple movable parts are connected together, each with their own degrees of freedom, then these degrees of freedom can work together additively to give the assembly a greater range of movement. Going back to the example of the arm in Figure 5-16, the rotational degree of freedom of the wrist, the hinge joint between the upper and lower arm, and the ball and socket joint between the arm and rotator means the grabber can reach and orient itself in many more possible ways than any of these degrees of freedom. For each type of translational and rotational movement, limits can also be set as to how far the movement can go. This might be a linear distance for a translation and number of degrees for a rotation. These limits are typically defined by the geometry of the joined parts (Figure 2-19). For the virtual assembly to behave like the real one, parts cannot pass through one another as though they were invisible! In many cases, geometry of parts is designed specifically to control both the number of degrees of freedom and the distance of movement of a moving part.

Figure 2-18 Joint Types
Joints are defined by the types of translational and rotational movements they can make.

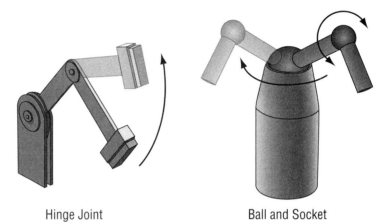

Hinge Joint   Ball and Socket

Figure 2-19 Movement Limits Many degrees of freedom have limits of movement determined by the geometry of the parts.

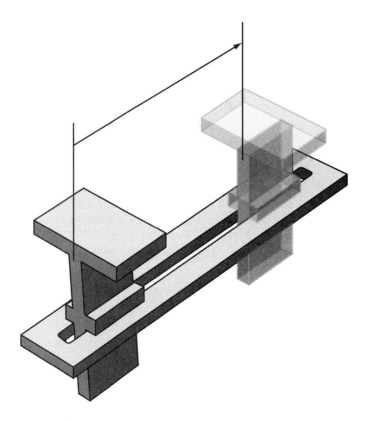

assembly	parent	joint
base part	subassembly	translation
hierarchical tree	fixed	rotation
children	movable	degrees of freedom

III. Student Materials

Use Project 3 Design Brief: Assembling a Medical Pump to complete the project.

UNIT 2: PROJECT 3   DESIGN BRIEF: ASSEMBLING A MEDICAL PUMP

Your medical technology company is about to begin the design of a new line of medical pumps to be used for infusing specially treated blood products into the bloodstream. As is the case with most pumps used in medical applications, this new line of pumps has to meet the following design criteria:

- The pumping action has to be gentle and not damage the fragile blood components (e.g., red blood cells).
- The areas of the pump that the blood products come in contact with have to be kept sterile and easy to clean.
- An accurate, constant flow of this blood product has to be infused into the bloodstream over a long period of time.

Three of the more common pump technologies are:

- Syringe pumps
- Diaphragm pumps
- Peristaltic pumps

1. In the previous project, you created the parts of one of these types of pumps. In this project, you will need to create an assembly of these parts.

2. Create an "exploded assembly" sketch showing how these parts will be assembled.

3. Next, create a hierarchical tree diagram showing how the parts in the assembly will be organized. On this hierarchy, label the types of joints between each parent-child pair of parts.

4. Finally, create an assembly model of the medical pump. Make sure you properly orient and locate each part. If your modeler allows it, define the appropriate degrees of freedom and movement limits.

PROJECT 4
Animation

I. Project Lesson Plan

1. Project Description

In this project, you will take the assembly created in Project 3 and animate it. The animation should depict the functionality of the syringe pump and also provide an opportunity to test whether designed part design geometry allows for the proper movement and functioning of the parts.

2. Learning Objectives

- You will learn the fundamental principles of 3D animation.
- You will be able to define the basic animation commands.
- You will learn how to apply animation principles to depict the functionality of an assembly.
- You will gain an appreciation of how animation can be used in the design and visualization process.

II. Background Information

1. Animation

Once a model is constructed, static images of the assembly model can be captured in different configurations. However, if you want to show your model in numerous configurations or show the model as it transforms from one configuration to another, then you will probably want to use **animation** tools to depict this. An animation is simply a series of static images strung together showing change in the scene over time. These images are called **frames**, with an animation typically having between 8 and 30 frames per second.

Since manually changing the scene for each frame would be very time consuming, animation packages provide tools for automating changes in the model. One process is called **keyframing**, or **tweening**. With keyframing, the animator determines what changes in the scene at specific, "key" frames of the animation (Figure 2-20). The software then determines how the scene will look as it transforms from one keyframe to the next, automatically generating the in-between frames (thus the name "tweening"). Typically, you will also indicate how many frames should be created between each keyframe, thereby controlling how fast the change will occur.

What the keyframes are and how they transition from one to another should be carefully planned out in advance. Professionals and students alike create **storyboards** to plan their keyframes and design their animations. A storyboard is nothing more than a series of framed sketches strung together much like a cartoon strip. Each frame depicts a keyframe of the animation with text and arrow notations indicating what will (and how it will) change between one keyframe and the next.

Another animation technique is called **path-based animation** (Figure 2-21). This tool is used heavily for controlling translation and rotation of the model. With this approach, a curved path line is laid out in space, along which the model moves. You can also specify how the model should rotate relative to the curving of the path line and, by indicating the number of frames for the path, how fast it should move along this line.

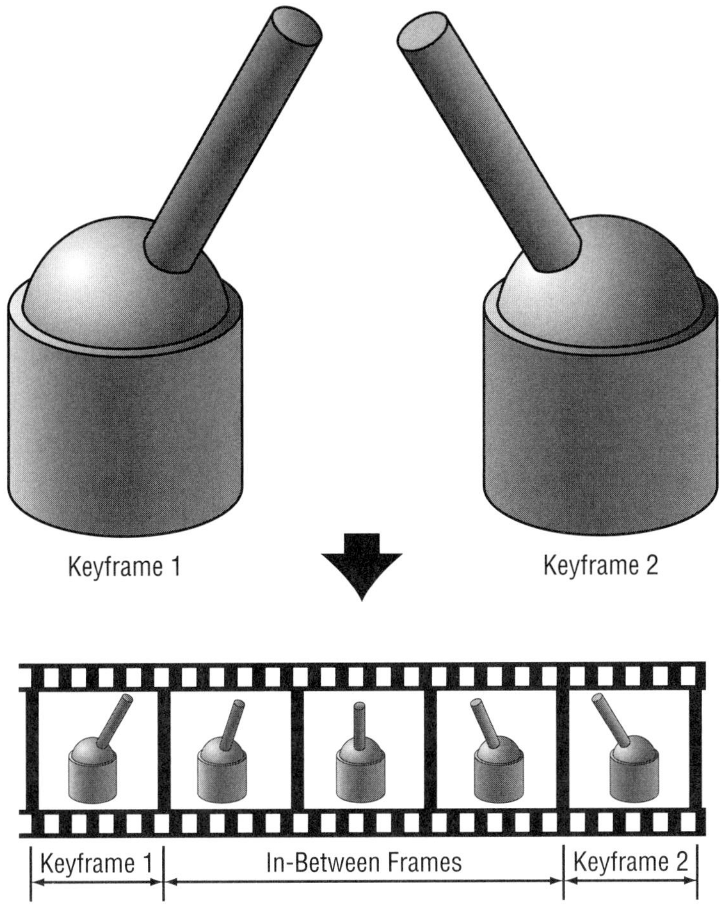

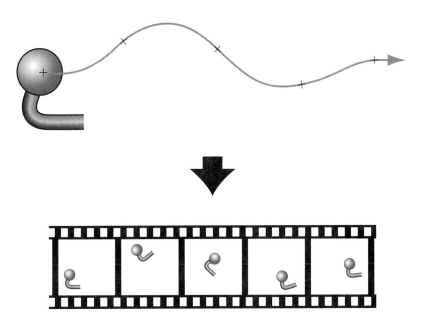

Figure 2-20 Keyframe Animation Keyframes define how parts should move in the in-between frames.

Keyframe 1

Keyframe 2

Keyframe 1 | In-Between Frames | Keyframe 2

Figure 2-21 Path-based Animation A path or vector describes how a part should translate and rotate over a series of frames.

With both keyframe and path-based animations, some animation packages allow the user to define changes that do not change uniformly over time. That is, you can have change happen at a slower or more accelerated pace at different points of an animation. A common technique is to use "ease-out" and "ease-in." These effects cause the part to move more slowly at first, accelerating to maximum speed in the middle of the animation section, and then slowing down again at the end.

One advantage of carefully creating an assembly of the product before animation is that the constraints you put into the assembly will limit the movements of the parts in the animation to only what is possible in the real product. If an articulating arm should rotate only 120 degrees, then you can move the arm to one extreme in one keyframe, to the other extreme in the next, and know that the resulting animation will show only allowable motion. Similarly, a well-constructed hierarchy will make it easier to define how parts can move relative to each other, creating more realistic and sophisticated movements.

Think back to the example of modeling the human body. Parts hierarchy and movement constraints mean that you don't have to carefully place and orient every finger and toe so that they stay "attached" in every keyframe or intermediate in-between frames. What the assembly model may not do, however, is keep independent parts from clashing (overlapping) with each other. You may have to track this by carefully reviewing frame sequences.

2. Types of Changes

Without change, there is no reason to do an animation. Your job as the animator is to decide what in the scene is going to change and when. These changes are either controlled frame by frame, or by techniques such as keyframing. If you have an assembly with multiple parts, there is a good chance that there will be parts that can change independently of each other. With parts that are hierarchically linked, you have to decide how each part will move to create the final effect you are trying to present. To create a successful animation, you must decide how to orchestrate these changes. Again, storyboards will be a key tool for helping to plan this out.

One of the most common types of change that occur in the scene of an animation are rigid transformations such as translational or rotational changes in parts. The same transformations that you used to place a part in an assembly can also be used to create change in the animation. In addition, there are also deformations of the geometry. In more sophisticated animation packages, these deformations can be controlled by the physical characteristics assigned to the object. For example, if you wanted to simulate a bouncing ball, the ball would be assigned a rubber-like material that would deform as it came in contact with a "rigid" surface.

Changes that would be considered part of the rendering process can also be altered over time in the animation. Here, one of the most common changes is in the viewpoint of the scene. In the animation, the camera can be moved to follow paths, much like a path-based animation of an object. Similarly, a camera may be "attached" to a moving object in the scene. Other changes in the camera also mimic the techniques used in film and television production. The camera can **zoom, pan**, or **tilt** at predetermined times or quickly switch between two or more cameras.

In addition, lighting can be changed in the animation. As with the camera, lights can translate and rotate to a point in a different direction. Lights can also change intensity, color, or focal angle (for spotlights). Finally, material characteristics of the objects can be altered over time.

Advanced lighting and rendering techniques are explored in more detail in **Project 5** of this unit.

Key Terms

animation	tweening	zoom
frame	storyboards	pan
keyframing	path-based animation	tilt

III. Student Materials

Use Project 4 Design Brief: Animating a Medical Pump to complete the project.

UNIT 2: PROJECT 4   DESIGN BRIEF: ANIMATING A MEDICAL PUMP

Your medical technology company is about to begin the design of a new line of medical pumps to be used for infusing specially treated blood products into the bloodstream. As is the case with most pumps used in medical applications, this new line of pumps has to meet the following design criteria:

- The pumping action has to be gentle and not damage the fragile blood components (e.g., red blood cells).

- The areas of the pump that the blood products come in contact with have to be kept sterile and easy to clean.

- An accurate, constant flow of this blood product has to be infused into the bloodstream over a long period of time.

Three of the more common pump technologies are:

- Syringe pumps
- Diaphragm pumps
- Peristaltic pumps

1. In the previous project, you created an assembly of one of these types of pumps. In this project, you will need to create an animation of this assembly to show how it functions.

2. Create storyboard sketches to show how you are going to animate the assembly.

3. Next, create an animation of the medical pump.

PROJECT 5
Rendering and Deformation

I. Project Lesson Plan

1. Project Description

In this project, you will explore rendering techniques to help visualize pump functionality. Specifically, you will look at how surface properties (e.g., color and reflectivity) and lighting are used to help communicate the pump's geometry and functionality. In addition to rendering, you will also study how geometry deformation plays a role in some pump technologies. You will learn about the relationship of general pump technologies and how deformable materials are used in different types of pumps. In this project, you will use modeling/animation deformation techniques to demonstrate these designs.

2. Learning Objectives

- You will learn the fundamental principles of 3D rendering.
- You will learn the fundamental principles of shape deformation.
- You will be able to define the basic rendering and deformation commands.
- You will learn how to apply deformation principles to depict the functionality of an assembly.
- You will gain an appreciation of how rendering can be used in the design and visualization process.

II. Background Information

1. Rendering

When creating a model, there is usually the goal of representing both the size and shape of the model and the **surface properties** of its parts. The surface properties of the model include its color, the texture of its surface, and how that surface responds to light striking it. How the surface looks to the viewer will also depend on the number, types, and locations of the lights shining on the object and the viewpoint of the viewer. The process of assigning surface properties, setting lights, and establishing a viewpoint is referred to as **rendering** the model.

a. Surface Properties

The most basic surface property is its **color**. Color can be thought of as having three basic components: hue, saturation, and value (or lightness). **Hue** is normally what we think of as the color. That is, red, green, or blue. **Saturation** is the amount of the dominant hue in relation to other hues that may be part of the color. As the saturation of a color increases, the color goes from a gray tone, through washed out pastels and on into rich, vibrant colors. The **value** (lightness) of a color is how light or dark a color is. In terms of rendering, value is determined by the amount of light striking that part of the model surface.

A simple rendered model may use a color of fixed hue, saturation, and value (Figure 2-22). However, this type of rendering will make the model look very "flat." Most rendering will vary the value of a surface based on its orientation to the one or more virtual light sources shining on the model (Figure 2-23). Faces of a part vary from each other not in their hue, but in how light or dark they are. The lightness of a face is determined by how each face is oriented relative to the light source. A face perpendicular (also called **normal**) to the direction the light is shining the lightest, while a face parallel to the light source (or pointing away from it) is the darkest.

The lightness and darkness of a face can vary based on a number of other rendering techniques. First, where shadows are cast can be calculated, darkening parts or all of a face. On the other side, reflected light from surfaces can be calculated and, where the light strikes, the surfaces lightened up. In most rendering systems, whole faces are not colored uniformly, but vary based on the shades of the neighboring faces. The renderer can also vary the coloring based on the quality of the surface and the type of light hitting the surface. A surface with high **specular reflection** will concentrate the light in highlights the way that glossy car paint does. In contrast, high **diffuse reflection** will diffuse light over a surface, much as velvet would.

Rendering systems can also support surface mapping. One type of mapping, **texture mapping**, is not really a texture so much as a swatch of patterned

Figure 2-22 Flat Shaded Model Rendering that does not vary in value makes the model look "flat."

Figure 2-23 Varying Value in the Shaded Model Varying the value across a surface or surfaces gives depth to the model.

color. Imagine a wallpaper pattern that is repeated over and over across the object. In defining the mapping, you would pick the swatch (pattern) you want to map with, then define its size, orientation, and how to stretch it across the surface. Another type of mapping, **bump mapping**, manipulates how light is reflected off the surface. Rather than gradually changing light reflection across a surface, bump mapping (as its name implies), allows regular small changes in light reflection to happen across the surface. Whereas texture mapping simulates surfaces that are not solid colors (e.g., marble or wood), bump mapping simulates unevenness in the surface, like a golf ball or corroded metal.

b. Lighting

Rendering of the surface of the model is intimately tied to the lighting used in the scene. Without light, no surfaces can be seen on the model. The most common light source is an **ambient light** source that provides an overall level of light for the entire scene. The effect is not unlike fluorescent lighting in a classroom. If the material on the model is a solid color, then all the surfaces of the model should be shaded exactly the same hue and value. While this type of light is useful to provide overall lighting, other types of lights should also be used to give depth and form to the model.

To differentiate between surfaces on the model, a light source with a particular directionality is needed. In addition, the renderer has to be capable of different values across the surface of the model to accurately simulate the light hitting the surface. A **point light** source imitates an incandescent lightbulb (Figure 2-24). This light source has a location but then shines a uniform amount of light in all directions. Because the light has a particular location, not all parts of the model get light hitting them at the same angle or at the same distance. For that reason, different surfaces will be lighter or darker than others.

Figure 2-24 Point Light Source A point light source behaves like an incandescent lightbulb.

A **spotlight** is like a point light source except that it has a direction and focal angle in addition to location (Figure 2-25). Just like its name, this type of light can be *pointed* at an object to highlight a particular part of the model. The **focal angle** determines whether a small or large part of the model is lit up. The focal angle creates a *cone of light* with the central axis of the cone being the direction of the spotlight. The larger the angle, the more it is like a standard point light source. Some spotlight sources can be designed to have their intensity drop off as you move to the outer edge of the focal angle or a farther distance to the object.

Finally, a **directional light** source acts like the sun (Figure 2-26). Here, all of the light rays are parallel to each other and at the same intensity no matter how far away the object is. The differing angles the surfaces are relative to the light rays shift how light or dark the surface is rendered.

Figure 2-25 Spotlight Source A spotlight is used to highlight selective parts of the model.

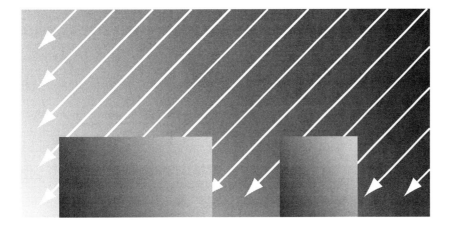

Figure 2-26 Directional Light Source A directional light is much like the light from the sun.

c. Viewpoint

Before the scene can be rendered, a **viewpoint** has to be established. You can think of setting a viewpoint as being similar to setting up a camera to take a picture. By altering the location, lens angle, and zoom on the camera, you can alter how much of the scene you can see and from what point in the scene. If you are creating an animation, you can think of your camera as being a camcorder. Just as in real life, you can also move your camcorder as you record. Viewpoint and camera control are discussed in more detail in **Project 6**.

2. Deformation

Surface deformation, or simply **deformation**, occurs when geometry has been altered by the interaction with other geometry. Deformation can also be achieved by the user manipulating local portions of the model geometry. This deformation can be permanent, or it can be temporary. For these temporary deformations, the surface returns to its original state after some period of time. Both permanent and temporary deformations are often part of an animation where deformations happen over time according to a predefined process. For example, other geometry (often invisible) is translated through space so that it interacts with the visible geometry, causing it to deform.

Imagine an invisible sphere moving into one that is visible and then moving back away. The viewer would see the side of the visible sphere deform inward and then reexpand. Deformations can be created between two visible objects by the laws of physics and material properties associated with the objects. Imagine now a rubber ball being dropped over a hard floor. The ball would fall, hit the floor, deform, bounce back up, and return to its original shape.

Deformation is one of many tools the modeler can use to model realistic properties of materials and objects. Many medical pump designs make use of the elastic nature of materials. For instance, diaphragm pumps make use of an elastic membrane to pull and push liquid through them. Peristaltic pumps squeeze down a flexible tube over its length to force liquid through it. For the modeler, the challenge is to determine how to realistically model these deformations.

Key Terms

surface properties	normal	point light
rendering	specular reflection	spotlight
color	diffuse reflection	focal angle
hue	texture mapping	directional light
saturation	bump mapping	viewpoint
value	ambient light	deformation

III. Student Materials

Use Project 5 Design Brief: Rendering and Deformation to complete the project.

UNIT 2: PROJECT 5   DESIGN BRIEF: RENDERING AND DEFORMATION

Your medical technology company is about to begin the design of a new line of medical pumps to be used for infusing specially treated blood products into the bloodstream. As is the case with most pumps used in medical applications, this new line of pumps has to meet the following design criteria:

- The pumping action has to be gentle and not damage the fragile blood components (e.g., red blood cells).

- The areas of the pump that the blood products come in contact with have to be kept sterile and easy to clean.

- An accurate, constant flow of this blood product has to be infused into the bloodstream over a long period of time.

Three of the more common pump technologies are:

- Syringe pumps
- Diaphragm pumps
- Peristaltic pumps

Part 1

1. In the previous project, you created an assembly of one of these types of pumps. In this project, you will need to render this assembly to help highlight its different parts.

2. Create three different renderings of your assembly. Choose colors, lights, and material properties that you think will help differentiate the different parts. These renderings will be used in advertisements and instruction booklets, so the users will need to clearly differentiate the different parts and have some idea what material the different parts of made of.

Part 2

1. Next, create an animation of either the diaphragm or peristaltic medical pump. Use one of the rendered models created in Part 1.

2. In the animation, deform the appropriate parts of the pump to demonstrate its functionality. You may have to make selected parts transparent or remove them for part of the animation to show the deforming part(s).

PROJECT 6
Camera Viewpoints and Particle Systems

I. Project Lesson Plan

1. Project Description

In this project, you will explore advanced modeling and animation techniques as a way of visualizing medical pump functionality. You can use alternate camera viewpoints to highlight particular parts of the pump or take the viewpoint of a blood cell going through the pump. Particle systems are an advanced modeling technique used to simulate the complex movement of fluids and gases and those elements that would move through them. For medical pumps, this might be the flow of blood cells through the pumps.

2. Learning Objectives

- You will learn to apply multiple cameras and viewpoints in a model scene.
- You will learn the fundamental principles of particle systems.
- You will gain an appreciation of how multiple viewpoints and particle systems can be used in the design and visualization process.

II. Background Information

1. Cameras

When you render your virtual model, you do so from a particular **viewpoint**. That is, the image you see on the computer is created by "looking" at your model and scene from a specific spot in the virtual world using viewing parameters that are not unlike what you use when taking pictures with a camera. Most modeling and animation software packages allow you to create one or more **virtual cameras** that define the viewpoint of the rendered scene.

a. Camera Lens

When defining the viewpoint, the first set of camera controls you should look at are those related to the virtual camera lens. Typically, the command names used by the software will parallel the terms used with real physical cameras. The **focal length** and **angle of view** (also called *field of view*) are related controls that achieve the same end (Figure 2-27). On a traditional camera, lenses of different lengths are used to capture images that take in larger or smaller amounts of the world in front of it. For example, a photographer might use a normal lens (around 50 to 55 mm) to capture what the human eye typically sees in some detail. How much you can see is measured by the angle of view. Hold your arms out in front of you with your hands together. Now start spreading them apart until you can no longer make out the details of your thumbs. The angle between your arms is the angle of view and this particular angle is about the angle of view of a normal lens. A **wide angle** lens (35 mm or less) takes in about the limit of what the eyes can see, but in more detail than we would get from simply looking straight ahead. A **telephoto** lens (80 to 300 mm) closes in

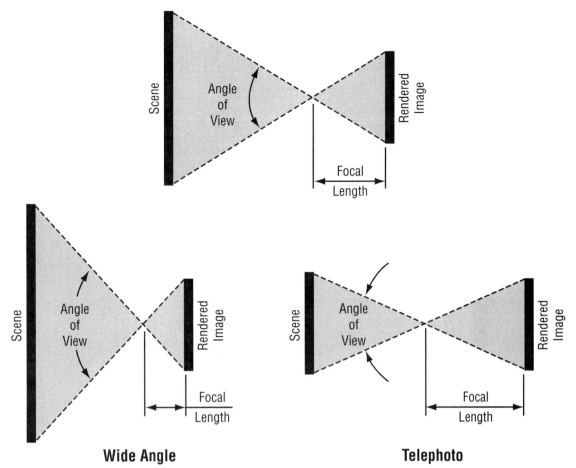

Wide Angle **Telephoto**

Figure 2-27 Focal Length and Angle of View of a Camera Lens A wide angle lens has a short focal length while a telephoto lens has a long focal length.

the angle of view to see only a small part of the eye's central view. A 200 mm lens would close down the field of view so your arms would only be about 10 degrees apart.

A common type of lens you would buy for a real camera would be a **zoom lens**. This type of lens allows you to change the focal length of the lens without changing lenses. When you "zoom in," you increase your focal length and narrow your angle of view. Zooming in gives the appearance of the objects in the scene getting larger. This happens because as the angle of view gets smaller, objects at the center of the scene expand to take up more of the total rendered image. Again, go back to using your arms in front of you and spread them to take in the width of, say, a door. The door might take up 10 degrees of your angle of view. If you were using a lens with a 50 mm focal length, this might only take up a quarter or so of your total angle of view. On the other hand, if you zoomed in to a 200 mm focal length, now that 10 degrees is taking up your entire view. Your virtual camera will also likely have a zoom function that behaves the same way.

With your virtual camera in your software, you might change your angle of view directly, or you may control it by changing the camera's focal length. In either case, the result will be the same. When the focal length increases on a traditional camera, the **depth of field** gets shallower (Figure 2-28). The depth of field indicates how much of the scene (going away from the camera) is in focus. The lens of a camera focuses on a particular distance away from the camera, but some of the scene in front or in back of the camera stays pretty much in focus. The shallower the depth of field, the less of the scene

Figure 2-28 Depth of Field A long focal length and a large aperture create a shallow depth of field.

Shallow

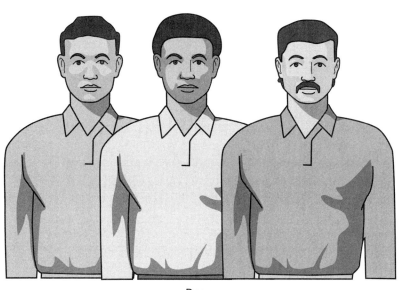

Deep

in front or behind the focal point stays in focus. So, with a telephoto lens, not only does the angle of view (side to side) get smaller, but the depth of field gets shallower. It makes the scene look "flatter."

Of course, it's the case with real lens—virtual cameras in the software do not have to follow this rule. If the software does not automatically adjust the depth of field when you change its focal length, you can often adjust it with an f/stop control. Real cameras use the f/stop to control the **aperture**: the size of the hole that lets light into the camera. The larger the aperture (smaller f/stop number), the shallower the depth of field. For example, an f/2 has a shallow depth of field while an f/11 has a much deeper depth of field.

b. Camera Movement

The previous section discussed ways of changing your view of the rendered scene by manipulating the camera lens. The other primary way you will change your view is by moving the camera. Once again, the terms used to describe these camera movements are taken from their real-world equivalents; in this case, cameras used in film and television.

If you imagine the camera mounted on a tripod, there are two major types of movements you can make: rotating the camera on the tripod while the tripod stays fixed to the ground or moving the whole tripod. For both these types of movements, thinking about how the camera movement relates to the direction the camera is pointing, that is, its **line of sight**, is also important.

The primary movement you would make while holding the tripod fixed to the ground and rotating the camera is a **pan** (Figure 2-29). Usually, a pan means rotating the camera horizontally as though the tripod were on a sidewalk and you were following a car coming down the street and then past you. However a pan might also go vertically, as though you were following a rocket launch. For a pan, the line of sight is always changing and all of the stationary objects in the scene seem to move and change angle in the rendered scene.

The other movements of the camera involve moving the (virtual) tripod. Imagine filming the car going down the street again, but this time you have the camera mounted in a car in the next lane over, moving at the same speed as the car you are filming. Now the car seems as though it is stationary while the background rushes by. This move, appropriately, is called a **truck** (Figure 2-30). With a truck movement, the camera moves perpendicular to the direction of the line of sight. With the camera still in the car, you can point it in the direction the car is moving and drive slowly toward a person standing in the middle of the street. The person will start small, but begin to fill up the field of view. This move, a **dolly**, is similar to a zoom, except you are not changing the focal length of your lens and, therefore, your depth of view. In this case, the person is getting larger in the rendered scene because you are actually getting closer to him or her. While a truck moves perpendicular to the line of sight, a dolly moves along the line of sight.

Another possible way is to move around an object. Once again, a person is standing in the middle of the road but now the camera tripod is on casters. The camera is wheeled in a circle around the person while keeping the camera pointed at the person. This move is called an **orbit** (Figure 2-31). Numerous other movements are possible by combining the previously mentioned movements and

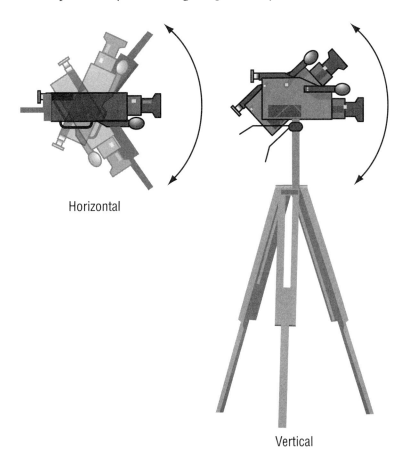

Horizontal

Vertical

Figure 2-29 Camera Pan Moves A pan rotates the camera while holding the tripod stationary.

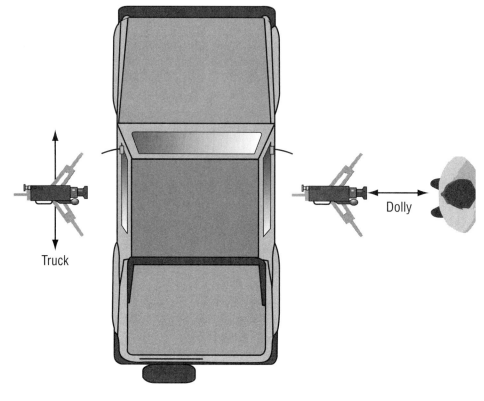

Figure 2-30 Camera Truck and Dolly Moves A truck moves perpendicular to the line of sight while a dolly moves with it.

Truck

Dolly

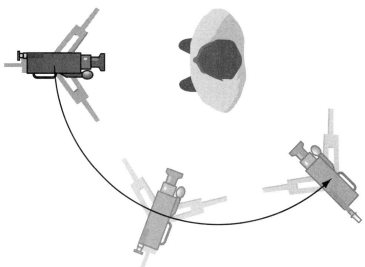

Figure 2-31 Camera Orbit Move An orbit moves around an object while holding the point of view on the object.

having the camera move off the ground. Any camera shot that takes you off the ground and films the scene at a height above a typical human view is called a **crane** shot. However, in many cases, you are creating animations of objects that seem to be floating in space. In these cases, the notion of a crane shot does not really apply.

In controlling these movements of the virtual camera, you can typically either program the movement of the camera directly or do it indirectly by tying the camera to another object in the scene. If the camera is supposed to simulate a driver looking out the front windshield of a car, then the easier way to control the camera is to fix its point of view relative to the car. Therefore, wherever the car moves, the camera will follow it appropriately.

Sometimes the camera movement is not that straightforward. Another alternative tool often available is to have the camera follow an invisible vector that you have drawn in the scene. In this case, you will have to define what the point of view will follow. For example, it might always stay perpendicular to the movement vector. Finally, much more complex camera movements can be created by moving the virtual tripod, panning, or changing the lens setting all at the same time. However, make sure whatever camera moves you plan serve the goal of more clearly demonstrating the functionality of the device you are visualizing.

c. Using Multiple Camera Views

An alternative to changing the settings on a single camera is to have multiple virtual cameras in the scene. Multiple cameras are useful when you have multiple key views of a model assembly that you want to capture. Rather than move a camera and lose its settings, you can create multiple cameras. As the model transforms based on the instructions you give it, you can "shoot" from different cameras at different times. Alternatively, you can also do what is done in traditional films and shoot multiple cameras at the same time (or reshoot the scene multiple times with different cameras). You can then go back and edit these different viewpoints into a final animation.

As with real film, multiple cameras should be planned out ahead of time. Think about what functionality you want to show in your model. Should you rotate the model, move the camera, or have multiple cameras shoot the model from different viewpoints? Planning and practice will help you make these decisions.

2. Particle Systems

Particle systems are a sophisticated type of procedural graphics. **Procedural graphics** include any type of transformation of the model (or camera) where the changes have been described as a series of procedures rather than defining exactly what the model should be doing in each frame. Imagine shooting a video of a person waving his or her arms. While you could pose the person, click one frame of the video, pose the person in the next incremental position, click one frame, and so on. Instead of doing that, you give the person an instruction (procedure) to wave his or her arms, then you shoot the video. A particle system is a way to simulate complex movement in your scene without having to micro-control all of the objects in the scene. With a particle system, you can have literally thousands of objects moving through your scene with only a small number of procedural settings.

There are many natural phenomena that you may want to simulate in your animation. This might include smoke rising or the flames that created the smoke. Similarly, you might want to simulate water spraying from a hose or, even better, particles in the water flowing down the hose. In the case of medical pumps, you might want to simulate blood cells flowing in and out of the pump as a way of demonstrating how the pump system makes fluid move through it. In all of these cases, particle system generators in more advanced modeling and animation systems can help you create this effect through procedural graphics.

Particle systems attempt to simulate how particles might behave in the real world. That is, how the physical laws would interact with particles. For that reason, you must describe the properties of the particle system for it to know how to simulate its behavior. For example, you would want to indicate whether the particles are large enough to be visible in the rendering, or so small they create a "cloud" like smoke would. Typically these particles would emit from some starting location. How many particles, how fast (their velocity), and in what initial direction they are moving are all parameters that might be specified.

Once the initial parameters are established, how they interact with the virtual world may be defined. For example, does gravity act on the particles? Similarly, are there model surfaces that the particles would bounce off of? Finally, further properties of the particles might be defined. If the particles are to change size, shape, or color over time, then these parameters might be added. More sophisticated systems may define other physical properties such as the elasticity of the particles or if they bounce off of each other. While particle systems are often used to create "fireworks" in an animation, the primary purpose should be to help visualize the physical properties of the model that would be hard to create with lower-level control of individual objects.

Key Terms

viewpoint	zoom lens	dolly
virtual camera	depth of field	orbit
focal length	aperture	particle system
angle of view	line of sight	procedural graphics
wide angle	pan	
telephoto	truck	

III. Student Materials

Use Project 6 Design Brief: Camera Viewpoints and Particle Systems to complete the project.

UNIT 2: PROJECT 6   DESIGN BRIEF: CAMERA VIEWPOINTS AND PARTICLE SYSTEMS

Your medical technology company is about to begin the design of a new line of medical pumps to be used for infusing specially treated blood products into the blood stream. As is the case with most pumps used in medical applications, this new line of pumps has to meet the following design criteria:

- The pumping action has to be gentle and not damage the fragile blood components (e.g., red blood cells).

- The areas of the pump that the blood products come in contact with have to be kept sterile and easy to clean.

- An accurate, constant flow of this blood product has to be infused into the bloodstream over a long period of time.

Three of the more common pump technologies are:

- Syringe pumps
- Diaphragm pumps
- Peristaltic pumps

Part 1

1. In the previous project, you created an assembly of one of these types of pumps. In this project, you will need to create an animation of this assembly to help highlight its functionality.

2. Create three different animations of your assembly in action. Choose different camera viewpoints and camera movements that you think will help highlight different parts of the pump and its functionality. These animations will be used in advertisements and instruction booklets.

Part 2

1. Next, create an animation of either the diaphragm or peristaltic medical pump. Use one of the rendered models created in Part 1.

2. In the animation, create a particle system depicting blood cells moving through your pump. You may have to make selected parts transparent or remove them for part of the animation to show the blood cell movement.

UNIT 3
Energy and Power Technology

Unit Overview

I. Introduction

Energy. It makes the world as we know it exist. Every aspect of our lives is intertwined with the availability of energy. What exactly is energy? What forms can it take? Is there a shortage of energy? These are often confusing issues for people to understand. Energy can be bought in the form of energy bars and drinks at your local grocery store, and your family most likely pays a substantial amount of money each month to your local utility for energy. Turn on the nightly news and you may hear of a looming energy crisis. What does all of this mean?

This unit will focus on developing a better understanding of energy and the technologies that allow society to harness energy for its uses. The introductory level projects focus on defining energy and its many forms. Starting with simple energy sources such as alkaline batteries and solar cells, you will have the chance to explore the role technology has played in converting less useful forms of energy to more useful forms. By designing an energy transfer device, you will explore the law of conservation of energy and energy transformations. The intermediate project focuses on renewable and nonrenewable energy resources and their associated

technologies that supply energy for automobiles, homes, and businesses. You will explore the role technology can play in conserving our nonrenewable resources and in successfully implementing renewable energy resources into everyday society. Throughout both levels, you will have the opportunity to explore societal and environmental issues relating to the use of energy and power technologies.

II. Unit Learning Goals

- You will develop an understanding of the historical perspective of specific energy and power technologies, including the key technological and scientific advancements and the people involved in the advancements of the particular tools.
- You will develop an understanding of the uses of various energy and power technologies and how they operate.
- You will develop an understanding of the societal implications of energy and power technologies.
- You will create visualizations to convey information about energy and power technologies.

III. Projects

Introductory Projects

PROJECT 1: History of Energy and Power Technologies

This project provides you with the opportunity to research the historical perspective of a particular technological tool that converts less useful energy forms to more useful energy forms. This project is useful in developing your understandings of the nature of technology and its relationship with society. You will research the history and background of a specific type of energy and power technology by using the Internet and other resources. This information may be communicated in the form of a timeline, drawings, posters, reports, or an electronic presentation as part of Project 2.

PROJECT 2: Technological Tools and Energy

You will research different technological tools that are commonly used in society to convert energy to more useful forms. These tools may include, but are not limited to, batteries and photovoltaic cells. You will then create conceptual visualizations that communicate how the technological tool converts the less useful energy into a more useful form of energy. This project is useful in developing an understanding of the role energy and power technologies play in providing our society with sources of useful energy forms.

PROJECT 3: Visualizing Energy Transfer Devices

You will design an energy transfer device that uses a series of energy transformations to complete a simple task. This project is useful in developing understandings of the forms of energy and the law of conservation of energy. The designs must be presented in a visual format and will indicate the flow of energy throughout the device.

PROJECT 4: The Environmental Impact of Batteries

You will research the impact of battery disposal on the environment. After performing the research, you will present the information in a way that will make the audience aware of these environmental impacts, as well as provide information on the proper disposal of batteries.

Intermediate Projects

PROJECT 5: Fuel Cell Technology and the Hydrogen Economy

You will learn about how hydrogen is used as an alternative source of energy. The way in which fuel cell cars operate will be examined. You will create a visual presentation that synthesizes your understanding of the workings of fuel cells. Creating the presentations will require researching the benefits and risks associated with this alternative source of energy. Also included in the presentation will be the current status of fuel cell technology in the marketplace as well as presentation of at least three sources for credible further information pertaining to fuel cell technology.

PROJECT 6: Small-scale Hydroelectric Power

You will explore hydroelectric power as a renewable energy source. In particular, you will develop an understanding of how the potential and kinetic energy of water can be converted into work. The project includes a brief history and description of hydroelectric power. You will examine and create visualizations of the main components of a typical hydroelectric system. This project culminates in the *Run-of-the-River* activity in which you conduct and report upon a Preliminary Feasibility Assessment of a potential small-scale or microhydropower source. You will also be asked to consider local requirements and restrictions, as well as the ecological and economical impact of building the system.

Advanced Project

You will complete an independent project through the use of visualization tools by researching a new topic dealing with energy and power technology or by expanding on topics covered in this unit. The objective of the advanced level is for you to further your skills in integrating research, problem solving through the design brief approach, and presentation. Your teacher will work with you to negotiate the topic, time allocated to the project, and design constraints.

IV. Unit Resources

The Resource index contains a listing of all resources associated with the unit. Included are relevant Web site links, books, and other publications. The Glossary provides definitions for all key terms listed in each project.

PROJECT 1
History of Energy and Power Technologies

I. Project Lesson Plan

1. Project Description

This project provides you with the opportunity to research the historical perspective of a particular technological tool that converts less useful energy forms to more useful energy forms. This project is useful in developing an understanding of the nature of technology and its relationship with society. You will research the history and background of a specific type of energy and power technology by using the Internet and other resources. This information may be communicated in the form of a timeline, drawings, posters, reports, or an electronic presentation as part of Project 2.

2. Learning Objectives

- You will research the historical background of a particular technological tool that converts less useful energy into more useful forms.

II. Background Information

1. Historical Perspective

Energy, or the ability to do work, has existed since the beginning of time. All organisms use various forms of energy to survive. For example, plants naturally undergo **photosynthesis**, which converts **solar energy** into food energy for the plant to use. Humans are also dependent on many forms of energy to live. However, it wasn't until the Stone Age, approximately some 25,000 years ago, that humans began to use particular forms of energy to improve the quality of life. Specifically, humans began to use fire, which is a form of **heat energy**, to cook food and to provide heat for shelter. Throughout history, people have used **kinetic energy** in the form of animals to assist with transportation.

Approximately 5000 years ago, Middle Eastern people began to use fire to melt tin and copper to form tools and weapons. They also harnessed **wind energy** as a means of transportation by using sailboats. At approximately the same time, **magnetic energy** was discovered. It is thought that this discovery most likely occurred in China. It didn't take long for humans to realize the potential of using magnetic energy to aid in identifying directions. Today, many people still use magnetic energy in the form of compasses to identify directions.

Almost 2500 years ago, the Greek scientist and philosopher Thales identified another form of energy. By rubbing a piece of amber with fur and noticing the effect of the amber on other objects, Thales identified **static electricity**.

Approximately 1000 years ago in China, people found black rocks that when burned produced more heat than wood. After visiting China in 1275, Marco Polo wrote a book describing what he had seen. After reading his book, Europeans began to pick up and burn the black rocks, which are now known as coal.

By the 1600s, Europeans had begun looking for their own supplies of coal to use for heating. Around the middle of this century, England was providing most of the coal for the world. Although coal was a great source of heat and it allowed wood to be used for building rather than burning, the coal business led to an increase in the amount of pollution in England. During this same century, Europeans began to use **solar heating**. Large **solariums** were built from glass to collect and store the sun's energy to help indoor plants grow during cooler weather.

The 1700s saw continuing advancements through the use of coal. The development of the steam engine allowed more coal to be mined and led to the ability to transfer **chemical energy** and heat energy to kinetic energy. In 1786, Luigi Galvani hung the legs of a dead frog on metal hooks during a thunderstorm to see if the lightning would make the legs twitch. The legs twitched before the storm began. Galvani thought that he had discovered a new form of **electricity**, which he called animal electricity (see Figure 3-1 for an illustration of Galvani's finding). Galvani's findings led to many scientists attempting to bring the dead back to life by electrifying them. This scientific belief was the basis for Mary Shelley's *Frankenstein,* in which the doctor's monster was brought to life by the electric shock from a lightning bolt. Not everyone believed in the idea of animal electricity, however. One such person was Alessandro Volta. During the 1790s, Volta discovered the truth: there was no such thing as animal electricity. The twitching of the frog's legs was the result of electricity formed from a chemical reaction between the metal railing and the metal hooks holding the legs. Based on his findings, Volta created the first **battery** in 1800 by placing alternate layers of copper and zinc in a jar of salt water. This battery became the first steady supply of electricity.

By the 1800s, humans appeared to have conquered energy. Technological advancements allowed society to harness multiple forms of energy and transform them into more useful forms of energy that improved the quality of life. In 1839, Edmund Becquerel first observed the **photoelectric effect**, which would eventually lead the way to the development of **photovoltaic cells**. Some of the first solar cells were built in the 1880s using selenium, but these cells were extremely inefficient.

In the early 1900s, Albert Einstein developed his scientific theories explaining the photoelectric effect. Armed with a better understanding of the science behind the photoelectric effect, scientists were able to begin building more efficient solar cells. While coal remained the major energy source, many developments were made during this time including the use of **hydroelectric power**, wind energy, and **geothermal energy** to generate electricity.

The twentieth century saw the invention of the automobile, which used energy from petroleum. It also brought the knowledge of how to create **nuclear energy**. As the twentieth century came to an end, people became more aware of the

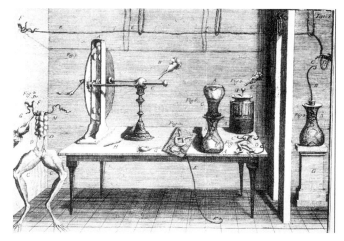

Figure 3-1 Galvani Believed He Discovered Animal Electricity Luigi Galvani mistakenly believed he discovered a new form of electricity.

(Courtesy of Smithsonian Institution Libraries; Washington, DC)

hazards and consequences of using coal, petroleum, and nuclear energy sources. This resulted in the development of new technologies that allow society to use alternative forms of energy that are renewable and have less negative impacts on the environment (see the **Resource Index** for Web sites with more information on the history of energy and renewable forms of energy).

Key Terms

energy	magnetic energy	battery
photosynthesis	static electricity	photoelectric effect
solar energy	solar heating	photovoltaic cells
heat energy	solarium	hydroelectric power
kinetic energy	chemical energy	geothermal energy
wind energy	electricity	nuclear energy

III. Student Materials

Use Project 1 Design Brief: Technological Tools and Energy Transformation to complete the project.

UNIT 3: PROJECT 1   DESIGN BRIEF: TECHNOLOGICAL TOOLS AND ENERGY TRANSFORMATION

You have been hired to work with a team of collections specialists at your local museum. Your team's assignment is to develop a collection of technological devices that were used to convert less useful forms of energy into more efficient forms. The collection must be representative of tools used in the past and present. The museum will be opening its "Technology of Power and Energy" collection in the fall. Your particular assignment is to research examples of technological tools/devices that are used for energy transformation and present your findings to the rest of the collections team so that they can decide whether to include it in the museum's collection.

Design Constraints (Things You Must Do)

- Select two technological tools/devices that are used for energy transformations. Be sure to include:
  - One tool/device from the past
  - One tool/device from the present
- Create a presentation about the tools/devices. Include in your presentation:
  - Inventor of the tool/device
  - How the tool/device works
  - What it is used for
  - Energy source and the type of output energy it converts to
  - Graphic of each device, which can be from the Internet, a freehand sketch, a computer-created illustration, or an animation
  - How the use of the technology has impacted society or the economy

PROJECT 2
Technological
Tools and Energy

I. Project Lesson Plan

1. Project Description

You will research different technological tools that are commonly used in society to convert energy to more useful forms. These tools may include, but are not limited to, batteries and photovoltaic cells. You will then create conceptual visualizations that communicate how the technological tool converts the less useful energy into a more useful form of energy. This project is useful in developing an understanding of the role energy and power technologies play in providing our society with sources of useful energy forms.

2. Learning Objectives

- You will identify and describe different forms of energy.

- You will explain how a technological tool converts a less useful form of energy into a more useful form of energy.

- You will communicate visually how technological tools allow society to use different forms of energy.

II. Background Information

1. Forms of Energy

Energy is defined as the ability to do work or transfer heat. There are many different forms of energy. All forms of energy can be classified as one of two categories of energy: **potential energy** or **kinetic energy**. Potential energy refers to stored energy or energy due to position (gravitational). Kinetic energy is energy due to motion.

Some of the most common forms of potential energy are **chemical energy**, stored **mechanical energy, nuclear energy**, and **gravitational energy**.

- Chemical energy refers to the energy stored in bonds between atoms and molecules. The energy in the bonds remains as potential energy until a chemical reaction occurs. During a chemical reaction, bonds are broken and new bonds are formed. During these processes, the stored chemical energy may be released or stored in new bonds. Examples of chemical energy include the energy found in petroleum, biomass, natural gas, and food. To release the stored chemical energy within these products, a chemical reaction, such as combustion or digestion, must occur.

- Stored mechanical energy refers to energy that is stored within an object as a result of a force applied to that object. For example, a stretched rubber band has stored mechanical energy as a result of the force that was applied to stretch the rubber band in the first place. Stretched bows and compressed springs also exhibit stored mechanical energy.

- Nuclear energy is the energy that holds the nucleus of an atom together. The mass of nuclear particles is converted to energy through the processes of fission and fusion.

- Gravitational energy is the energy an object has due to its position. Examples of gravitational energy include a ball resting at the top of an incline and the water stored in the reservoir of a hydropower dam.

Kinetic energy also has several forms of energy, which include **motion, sound, radiant energy, thermal energy**, and **electrical energy**. As discussed earlier, kinetic energy is energy due to motion.

- Motion or mechanical energy refers to the movement of objects as a result of an applied force. Wind and hydropower are both examples of motion or mechanical energy.
- Sound energy refers specifically to the movement of energy through longitudinal waves, called sound waves. A longitudinal wave differs from a transverse wave. A transverse wave is a typical wave with a peak and trough. Longitudinal waves are compression waves, meaning that the medium is compressed and then uncompressed as the energy moves through the medium. Sound energy is created by the application of a force to a material, which causes the material to vibrate. The energy is then moved throughout the material through the longitudinal wave.
- Radiant energy is the movement of electromagnetic energy by transverse waves. Examples of radiant energy include light, x-rays, gamma rays, radio waves, and **solar energy**.
- Thermal energy is commonly referred to as heat. It is the internal energy of a substance caused by the vibration and movement of atoms and molecules within that substance. **Geothermal energy** is a type of thermal energy.
- Electrical energy is the movement of electrons throughout a substance. Lightning and **electricity** are common forms of electrical energy.

See the **Resources Index** for Web sites with more information on energy.

You may think that energy is energy, but this is incorrect. Different energy sources may be more or less useful to our society. This is determined by the quality of the energy. Energy quality refers to the measured ability of a particular energy source to do useful work. Energy that is high quality is very concentrated and can perform lots of useful work. Examples of high quality or more useful energy include electricity, coal, and nuclear energy. Low quality energy sources are disorganized and can perform very little useful work. Examples of lower quality energy sources include normal sunlight, wind at moderate speeds, and geothermal energy sources.

For this project, you are to create visualizations that explain how technology has converted a less useful form of energy to a more useful form of energy. For example, solar energy exists as long as the sun continues to burn. However, the quality of this energy is fairly low and, therefore, less useful to society without the aid of a **photovoltaic cell**. The technology involved in the photovoltaic cell transforms the solar energy into electricity, a higher quality and more useful form of energy that helps to heat homes and power automobiles, radios, and calculators.

Another example of technology transforming a low quality energy source into a higher quality energy source is a **battery**. The low quality energy source within the battery is in the form of stored chemical energy. Without the other component of the battery, this less useful form of energy would be of little interest to society. However, when technology is applied, this low quality energy source is transformed into a much higher quality source known as electricity.

2. How Batteries Work

The common household battery is frequently known as a dry cell or an alkaline cell battery. It is comprised of several basic components. See Figure 3-2 for a diagram of the components. To understand how a battery works, you must understand that the

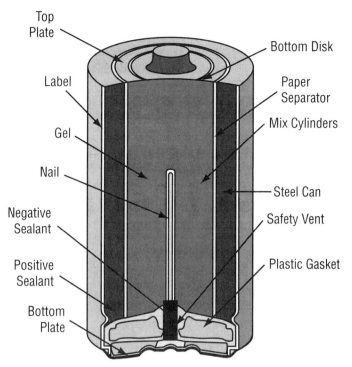

Top
Plate

Label

Gel

Nail

Negative
Sealant

Positive
Sealant

Bottom
Plate

Bottom Disk

Paper
Separator

Mix Cylinders

Steel Can

Safety Vent

Plastic Gasket

Figure 3-2 A Battery Converts Stored Chemical Energy into Electricity A battery transforms a low quality energy source into a higher quality energy source.

(Courtesy of Rayovac.com)

source of energy for the battery is an electrochemical reaction. An electrochemical reaction involves the transfer of electrons from one chemical reactant to another. These types of reactions are commonly called **redox** or **oxidation–reduction** reactions. The chemical that loses an electron to the other substance is oxidized. The chemical that gains or accepts the electron is reduced. For an electrochemical reaction to take place, both oxidation and reduction must occur.

All batteries have two terminals, an **anode** and a **cathode**. These terminals are the **electrodes** where oxidation and reduction occur. The positive terminal is known as the cathode. This is the electrode at which reduction occurs. The anode is the negative terminal and is the site of oxidation, or the loss of electrons.

There are several mnemonic devices that will help you to remember the difference.

- *LEO the lion goes GER.* LEO means "loses electrons oxidized" and GER means "gains electrons reduced."
- *OIL RIG.* OIL stands for "oxidation is losing" and RIG stands for "reduction is gaining."

To remember which process is occurring at the terminals, use the following phrases:

- *RED CAT.* This stands for *red*uction occurs at the *cat*hode
- *AN OX.* This stands for *ox*idation occurs at the *an*ode.

Different battery types are based on different electrochemical reactions. To have a different reaction, the materials used for the anode, cathode, and electrolyte must change. The battery system described in detail in the following text is an alkaline battery.

The entire reaction takes place within the container. The container is a steel can in which the materials that make up the cathode are placed. The cathode is generally molded to the inside of the container wall and consists of a manganese dioxide mixture and carbon. A special fabric called the separator is then placed in the container to separate the two electrodes. The anode is composed of powdered zinc metal and formed into a disk. An important piece of the battery is the electrolyte solution. This solution usually consists of potassium hydroxide and water. It is placed within the container and facilitates the movement of ions within the battery. The final component of the battery is the collector. The collector is a brass pin that is inserted in the middle of the cell and conducts electricity to the outside current.

Combining all of the component pieces of the battery is the technological advancement that has allowed our society to convert the less useful chemical energy into a more useful electrical energy. Confining the reaction within the container and controlling it through the use of a separator transform the once low quality chemical energy into higher quality electricity.

See the **Resources Index** for Web sites with more information on batteries.

3. Types of Batteries

There are many different types of batteries. The major difference between the battery types is the electrochemical reaction that occurs within the cell. Table 3-1 and Table 3-2 list some of the most common types of batteries and the materials used for the anode, cathode, and electrolytes. They also list some of the more common applications for each particular battery type. The materials used for the anode, cathode, and electrolytes dictate the type of electrochemical reaction that occurs within the battery. Table 3-1 focuses on primary batteries, or batteries that are non-rechargeable. Table 3-2 describes secondary or rechargeable batteries.

4. Solar Cells

Solar cells or photovoltaic cells convert sunlight, a form of radiant energy, into electricity. The basic components of a solar cell include:

- Two semiconductor layers
- Back and front contact
- Cover
- Antireflective coating

Table 3-1 Common Primary Battery Types

Common primary, or non-rechargeable, battery types and their applications are listed in the table. The materials used for the anode, cathode, and electrolytes dictate the type of electrochemical reaction that occurs within the battery.

Battery Type	Anode	Cathode	Electrolyte	Applications
Zinc carbon (dry cell)	Zinc	Manganese dioxide	Ammonium chloride or zinc chloride solution	Flashlights, toys, many other common household items
Alkaline cells	Zinc powder	Manganese dioxide powder	Potassium hydroxide solution	Radios, photo-flash applications, watches
Mercury oxide	Zinc or cadmium	Mercuric oxide	Potassium hydroxide	Small electronic equipment, hearing aids, photography, alarm systems, emergency beacons, detonators, radio microphones
Zinc air	Zinc powder and electrolyte	Oxygen	Potassium hydroxide	Hearing aids, pagers, electric vehicles
Lithium	Lithium	A variety of materials (lithium cells may have either liquid or solid cathode materials)	A variety of organic liquids (some lithium batteries have solid electrolytes)	Pacemakers, defibrillators, watches, meters, cameras, calculators

Table 3-2 Common Secondary Battery Types

Common secondary, or rechargeable, battery types and their applications are listed. The materials used for the anode, cathode, and electrolytes dictate the type of electrochemical reaction that occurs within the battery.

Battery Type	Anode	Cathode	Electrolyte	Applications
Lead acid	Sponge metallic lead	Lead oxide	Dilute sulfuric acid solution	Automobiles, construction equipment, recreational vehicles, standby/backup systems
Nickel cadmium	Cadmium	Nickel oxyhydroxide	Potassium hydroxide solution	Calculators, digital cameras, pagers, laptops, tape recorders, flashlights, medical devices, electric vehicles, space applications
Nickel metal hydride	Rare earth or nickel alloys with many metals	Nickel oxyhydroxide	Potassium hydroxide	Cellular phones, camcorders, emergency backup lighting, power tools, laptops, portable electric vehicles
Sodium sulfur	Molten sodium	Molten sulfur	Solid ceramic beta alumina	Electric vehicles, satellites
Lithium ion	Carbon compound, graphite	Lithium oxide	Solutions of lithium compounds	Laptops, cellular phones, electric vehicles
Alkaline manganese	Zinc	Manganese dioxide	Potassium hydroxide solution	Consumer devices
Nickel zinc	Zinc	Nickel oxide	Potassium hydroxide solution	Electric vehicles, standby load service

Two of the easiest components to understand are the cover and the antireflective coating. The cover, which can be made of glass or other materials, protects the solar cell from exposure to the weather. The antireflective coating is applied on top of the solar cell before attaching the cover. The purpose of this coating is to keep the solar cell from reflecting any of the sunlight away from the cell. If this coating were not applied, then the cells would be much more inefficient as a result of reflecting the sunlight away from the cell before it had a chance to interact with the cell to produce electricity. See Figure 3-3 for a representation of a solar cell.

The major components of a solar cell are the two semiconductor layers (see Figure 3-3). There are a number of materials that are suitable for use in the development of the semiconductor layers. There is no one ideal material for use in semiconductors, but silicon is the most commonly used material.

An interesting thing about semiconductors is that they can be treated with different substances to become positive or negative. This process of intentionally adding a substance is known as **doping**. In preparing silicon for a solar cell semiconductor, two doping processes must occur. N-type silicon is prepared by doping the silicon with phosphorous. The *N* stands for negative because the resulting semiconductor has a prevalence of free electrons. P-type silicon is prepared by

Figure 3-3 Solar Cell
The components of solar cells allow them to efficiently convert sunlight into electricity.

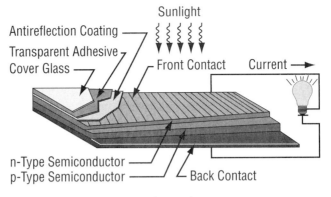

(Courtesy of U.S. Department of Energy)

Figure 3-4 N- and P-layers of a Photovoltaic Cell An electric field is created at the junction of the P-layer and N-layer semiconductors.

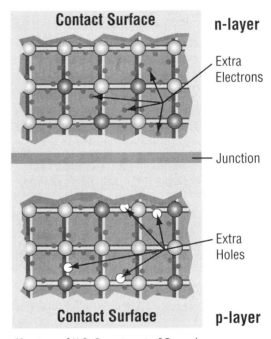

(Courtesy of U.S. Department of Energy)

doping the silicon with boron. The *P* stands for positive and refers to the free holes left by the absence of electrons.

By themselves the N- and P-type semiconductors are unexciting, but when they are placed in contact with each other, the energy can begin to flow. An electric field is created at the junction of these two semiconductors. The free electrons on the n-type silicon are looking for free holes, which are found on the p-type silicon. See Figure 3-4 for a representation of this process. Eventually, the free electrons and free holes reach equilibrium. When this occurs, the electric field acts as a diode. This means that the electrons are allowed to move only from the P side to the N side, not the other way around.

Now that we have a basic understanding of how the semiconductor layers within the cell work, let's look at the role that sunlight plays in making this solar cell work. As radiant energy from the sun hits the cell, some of the light causes the electrons within the semiconductors to move. This movement creates an electric current. Not all of the sunlight is capable of causing the electrons to move. Each semiconductor has an energy band gap that determines which wavelengths of the light spectrum have enough energy to cause the movement of electrons. Only the light energy within the band gap can result in the movement of electrons.

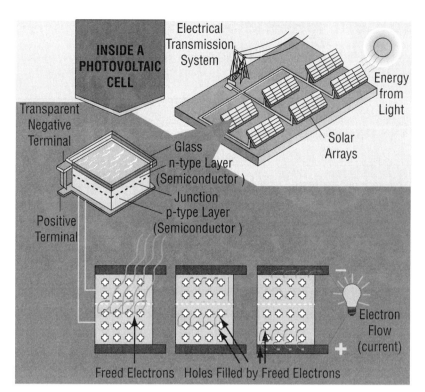

(Courtesy of U.S. Department of Energy)

Figure 3-5 Inside a Photovoltaic Cell Sunlight causes the electrons within the semiconductors to move. This movement creates an electric current.

Light energy with a wavelength smaller than the band gap does not have enough energy to move an electron and thereby creates a free hole. This light energy passes through the cell as if it were transparent. Light energy with a wavelength greater than the band gap has enough energy to move electrons and create free holes. Any excess energy is lost. Therefore, the energy of the sunlight is necessary to cause the electrons to move. See Figure 3-5 for a representation of this process.

The solar cell also consists of a front and back contact. These contacts allow the electrons to enter the circuit and complete the electrical circuit. An external circuit is needed to allow the electrons to move from one side of the cell to the other. The bottom or back contact is usually made of metal, which is good for conduction. However, the front contact cannot be a solid piece of metal because it would block the light from hitting the semiconductor. Therefore, a contact grid made of metal is used to allow the electrons to enter the circuit. Even by using a grid rather than a solid piece of metal, some of the photons from the light are blocked from the semiconductor layer.

See the **Resources Index** for Web sites with more information on photovoltaic cells.

Key Terms

energy	radiant energy	redox (oxidation–
potential energy	thermal energy	reduction)
kinetic energy	electrical energy	anode
chemical energy	solar energy	cathode
mechanical energy	geothermal energy	electrodes
gravitational energy	electricity	doping
motion	photovoltaic cell	nuclear energy
sound	battery	

III. Student Materials

Use Project 2 Design Brief: Technological Tools and Energy to complete the project.

UNIT 3: PROJECT 2 DESIGN BRIEF: TECHNOLOGICAL TOOLS AND ENERGY

You and your classmates have been invited by a science teacher in your school to make a presentation to a science class that is currently studying energy and its transformations. The science teacher has asked that you make a presentation explaining how technology has impacted society's ability to convert less useful energy sources to more useful energy sources. As your audience is currently studying energy and its transformations, you will also want to address the science of the energy transformation caused by the technological tool.

Design Constraints (Things You Must Do)

Your presentation must include a visualization that explains how a specific technological tool converts a less useful form of energy into a more useful form of energy. The visualization must focus on only one type of technological tool. Your visualization will also include an explanation of the types of energy being used and transferred by the technological tool.

PROJECT 3
Visualizing Energy Transfer Devices

I. Project Lesson Plan

1. Project Description

You will design an energy transfer device that uses a series of energy transformations to complete a simple task. This project is useful in developing understanding of the forms of energy and the law of conservation of energy. The designs must be presented in a visual format and will indicate the flow of energy throughout the device.

2. Learning Objectives

- You will understand why energy is neither created nor destroyed, by using the law of conservation of energy.
- You will design an energy transfer device that illustrates at least three transformations of energy to complete a specified task.

II. Background Information

In Project 2, you learned about the many forms of **energy** and how technology has allowed our society to change low quality energy into higher quality energy. In this project, you will learn about an important scientific principle: the **law of conservation of energy**. This law states that energy can neither be created nor destroyed, but it can change forms. This means that the amount of energy present in the universe remains constant; however, the form of the universe's energy is constantly changing.

Given the examples used in **Project 2**, we see that a **battery** does not create **electrical energy**. It actually transforms stored **chemical energy** into electrical energy. The same is true with a **photovoltaic cell**. The cell transforms **solar energy** into the more useful electrical energy. Therefore, it is conceivable that we could describe the role of technology in energy sources, such as batteries and solar cells, as transformers of energy.

To understand the law of conservation of energy, you must realize that energy may be transformed into more than one type of energy. The total amount of energy in all of the energy types present at the end of the transformation must equal the total amount of energy present before the transformation. A good example of this is the common light bulb. In a light bulb, electrical energy is converted into light energy. However, all of the electrical energy going into the light bulb is not converted into light energy. In fact, most of the electrical energy input is actually transformed into **heat energy**. To comply with the law of conservation of energy, the total amount of electrical energy put into the light bulb must equal the sum of the heat and light energy released by the light bulb.

As you design your energy transfer devices, you will need to think about all of the forms of energy that are being transformed and how to code these energy transformations in your visualizations.

energy

law of conservation of
energy

battery

electrical energy

chemical energy

photovoltaic cell

solar energy

heat energy

III. Student Materials

Use Project 3 Design Brief: Visualizing Energy Transfer Devices to complete the project.

UNIT 3: PROJECT 3   DESIGN BRIEF: VISUALIZING ENERGY TRANSFER DEVICES

In this project, you and your teammates will design an energy transfer device that will release a golf ball from a specified starting point and end by having the golf ball fall into a cup at a specified endpoint. This device will use a minimum of three different types of energy to complete the specified task.

Your final product will be a visualization of your design that shows the flow of energy throughout the device.

Design Constraints (Things You Must Do)

- Your design must include at least three different types of energy. Some examples are electrical, fluid, chemical, and mechanical sources.

- Your device must be able to fit within a two-foot cube when built.

- The device must not weigh more than 45 pounds when built.

- The golf ball must begin at the release point at the top of the cube and must be moved by a mechanical release.

- The ball must continue moving at all times.

- The ball completes the task by falling into a cup at the ending point located at the bottom of the cube.

- Your visualization must indicate the route of the golf ball and the type of energy being used at each point along the route. You must also identify the points at which energy transformations are occurring and what types of transformations are occurring.

- Based on the law of conservation of energy, energy can be neither created nor destroyed. Your visualization must account for all types of energy to follow the law of conservation of energy.

PROJECT 4
The Environmental Impacts of Batteries

I. Project Lesson Plan

1. Project Description

You will research the impacts of battery disposal on the environment. After performing the research, you will present the information in a way that will make the audience aware of these environmental impacts as well as provide information on the proper disposal of batteries.

2. Learning Objectives

- You will identify possible environmental impacts of improper battery disposal methods.
- You will explain the proper methods of battery disposal.
- You will create a presentation that will educate the general public on the environmental impacts of improper battery disposal, as well as the proper disposal methods.

II. Background Information

1. Alkaline Batteries

While **batteries** make our lives much easier and definitely more mobile, the use and disposal of batteries involves serious consequences that must be considered. Battery technology has greatly improved over the past few decades. The advent of rechargeable batteries has done much to reduce the number of batteries that must be disposed of. However, it is common to find many households using primary or nonrechargeable batteries because they are slightly cheaper. It is estimated that Americans dispose of 84,000 tons of **alkaline batteries** in their household trash each year. This represents approximately 20 percent of all hazardous materials found in landfills across the United States. While many states and localities do not have regulations regarding the disposal of primary batteries, it is important for our society to consider the possible environmental impacts of the disposal of these batteries and other battery types.

During the 1990s, legislation was passed in the United States that regulated the use of mercury in alkaline batteries. However, these and other primary batteries may have other environmentally toxic metals in them. Under normal conditions, an alkaline battery presents little to no environmental hazard. However, once that battery is in the landfill, the conditions are no longer normal. The battery may become damaged as the result of physical damage (e.g., crushing or cutting) or degradation from moisture, heat, or other materials with which it is in contact. If this happens, little can be done to prevent the leakage of hazardous materials into the soil.

Dumping batteries in the landfill is not the only problem. Batteries that are incinerated can release harmful vapors into the air. One of the simplest ways to reduce the possible environmental impacts of alkaline batteries is to invest in rechargeable batteries. These rechargeable batteries will eventually have to be disposed of, but they will represent a much smaller quantity than those of primary batteries.

2. Lead Acid Battery Disposal

The disposal of **lead acid batteries** presents some of the most dangerous environmental and health threats. The two main components of lead acid batteries are lead and sulfuric acid. Research has linked lead exposure to central nervous system damage. Located in the supplementary materials for this project are several Web sites and news articles containing information regarding the disposal of lead acid batteries (see **Resource Index**). This information demonstrates that simply burying lead acid batteries is unsafe. While it is understandable that lead acid batteries may have been buried decades before the environmental impacts were known, it is shocking to find that even today batteries are discarded on the side of the road and in the woods. Figure 3-6 is a photograph taken at the time of writing these materials. The picture was taken right beside the athletic fields of a public high school. If you examine the image closely, you will notice that the battery is beginning to corrode.

So how exactly do you get rid of that old lead acid battery? It's actually quite simple. When you purchase a new lead acid battery, the retail outlet is responsible for collecting your old battery. The collected batteries are then recycled. The recycling process removes the lead and other materials so that they can be reused in making new batteries. The sulfuric acid is recovered from the battery and neutralized so that it may be discarded of safely. By no means should old lead acid batteries simply be thrown out in the woods. Exposure to the weather will cause the batteries to corrode. Eventually, both the lead and the sulfuric acid will leach into the ground and possibly reach the local water supply.

(Dan Sullivan/Alamy)

(Colin Underhill/Alamy)

Figure 3-6 Discarded Lead Acid Batteries Notice how the battery on the right is beginning to show signs of corrosion.

3. Other Battery Disposal

Nickel cadmium batteries and **nickel hydride batteries** also present possible environmental and health threats. The metal component (e.g., lead, mercury, nickel, cadmium) of these batteries is often referred to as **heavy metals**. This popular term is often used to refer to particularly dense metal elements that are often harmful to the environment. If disposed of in a landfill, these heavy metals can leak into the soil and contaminate water supplies as well. Many manufacturers now offer programs in which they will accept your old nickel cadmium and nickel hydride batteries when you purchase a new battery. If the manufacturer does not have this option, you should contact your local solid waste department to find out how your town handles the disposal of these batteries.

Key Terms

batteries

alkaline batteries

lead acid batteries

nickel cadmium batteries

nickel hydride batteries

heavy metals

III. Student Materials

Use Project 4 Design Brief: The Environmental Impact of Batteries to complete the project.

UNIT 3: PROJECT 4   DESIGN BRIEF: THE ENVIRONMENTAL IMPACT OF BATTERIES

As an employee of the Solid Waste Department of your state, you have been asked to develop informational materials to launch a new campaign on the disposal of batteries. Your supervisor has asked that the materials educate the general public on the possible environmental impacts of improper battery disposal, in addition to the proper methods of battery disposal. Your supervisor has given you the freedom to choose whether your campaign will consist of brochures, television commercials, or Web pages.

Design Constraints (Things You Must Do)

- Present a graphic description of the components of the most common types of batteries.
- Emphasize the possible environmental problems caused by the improper disposal of batteries.
- Include guidelines for the proper disposal of both primary and secondary batteries.

PROJECT 5
Fuel Cell Technology and the Hydrogen Economy

I. Project Lesson Plan

1. Project Description

You will learn about how hydrogen is used as an alternative source of energy. The way in which fuel cell cars operate will be examined. You will create a visual presentation that synthesizes your understanding of the workings of fuel cells. Creating the presentations will require researching the benefits and risks associated with this alternative source of energy. Also included in the presentation will be the current status of fuel cell technology in the marketplace as well as at least three sources for credible further information pertaining to fuel cell technology.

2. Learning Objectives

- You will learn about fuel cells and how they produce electricity used to operate vehicles.
- You will learn about the benefits of fuel cells as well as any risks or limitations of this type of technology.
- You will develop your research skills by locating, reading, and evaluating additional sources of reliable information pertaining to fuel cell technology.

II. Background Information

1. Sources of Energy

The one thing that almost all economies throughout the world have in common is that they are extremely dependent on **energy** sources. Energy sources sustain businesses and are at the heart of manufacturing and transportation of goods and services needed across the globe. The energy industry is of vital importance, which is vividly seen when the demand exceeds the supply. Think about what happens when cities experience a blackout. Depending on what the weather situation is, not having **electricity** can wreak havoc. People experience heat stroke or conversely hypothermia. Restaurants cannot cook and serve food so their inventory becomes spoiled or rotten, and they lose any profit they would have normally made. Movie theaters are empty, as are sport arenas and shopping malls.

It is apparent that power and energy drive the economy. In 2006 alone, end-users (including Transportation, Industrial, Commercial, and Residential customers) spent over $1 trillion on energy, according to the U.S. Dept. of Energy. Figure 3-7 shows the average cost of **fossil fuels** at United States electric utilities for years 2005 and 2006.

There are many sources of energy. **Fossil fuels** are the most commonly used, representing approximately 85 percent of energy consumption in the United States. See Figure 3-8 for depictions of some common fossil fuels. Fossil fuels are things like oil, coal, and natural gas. Fossil fuels aren't the only sources of energy; other types of

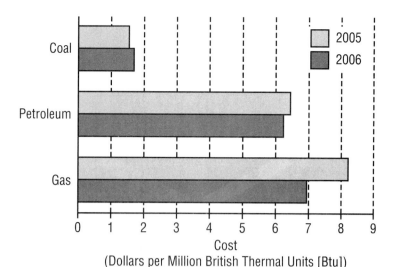

Figure 3-7 Average Cost of Fossil Fuels at U.S. Utilities in 2005 and 2006

Cost
(Dollars per Million British Thermal Units [Btu])

(Courtesy of U.S. Department of Energy)

Figure 3-8 Fossil Fuels Are Used to Sustain Global Economies The economy is dependent on the use of fossil fuels. As the rate of consumption continues to increase, many are concerned that the supply will become depleted.

(top left: Tatiana777, 2008/Shutterstock.com; top right: Brad Remy, 2008/Shutterstock.com; background: 2happy, 2008/Shutterstock.com; bottom left: Dmitri Mikitenko/Fotolia; bottom right: Grapheast/Alamy)

energy sources include **hydroelectric** (water), **wind, nuclear,** and **geothermal**. With the need for energy sources growing and the cost of the resources growing, efforts are being made to find alternative sources of energy. The hope is that alternative sources of energy will be more cost effective and produce less pollution.

Alternative sources of energy include **solar energy** and **hydrogen energy** (see the **Resources Index** for Web sites with more information on alternative sources of energy). Hydrogen energy has dramatic growth potential and is expected to develop and mature over the next 25 to 50 years. There are already products designed to make use of hydrogen energy. **Fuel cell cars** are one example. This project will focus on fuel cell cars and examine how the technology works, as well as the benefits and risks of **fuel cells**.

2. Fuel Cell Technology: An Alternative Energy Source

Fuel cells represent one answer to the challenge of finding energy sources that are reliable and clean. Car manufacturers are designing and developing cars that are powered by fuel cells. Fuel cells use hydrogen and oxygen to produce electricity, which is used to power the motor.

So how do fuel cells produce electricity? The process starts out as **chemical energy** is converted into electricity. Rather than using one fuel cell, fuel cell cars actually combine several single fuel cells into stacks. The reason individual fuel cells must be combined into a stack is so that they can produce enough electrical power to operate the vehicle. The more power needed, the more cells in the stack. To understand this process, let's first look at a single fuel cell.

An individual fuel cell consists of two flow field plates separated by a **membrane electrode assembly (MEA)**. The MEA itself consists of several components. In the center of the MEA is the **proton exchange membrane (PEM)**. On each side of the PEM is a thin layer of platinum catalyst. The MEA also consists of two **electrodes**: an **anode** and a **cathode**. The anode is located on one side of the PEM and is separated by the catalyst. The cathode is located on the opposite side of the PEM and is also separated from the PEM by the catalyst. Figure 3-9 show the components of a fuel cell.

Now that we are familiar with the components of the fuel cell, let's look at the actual process that occurs inside the fuel cell to produce energy. Basically, a fuel cell produces energy by converting chemical energy into electricity. The chemical energy comes from combining oxygen from the air with hydrogen gas. The flow field plates are used to direct the hydrogen and the oxygen to the appropriate parts of the fuel cell. Hydrogen from the onboard storage tank is supplied to the anode, and oxygen from the air is directed to the cathode.

The platinum catalyst alongside the anode promotes the separation of hydrogen into protons and electrons. The protons move from the anode to the cathode by passing through the proton exchange membrane (PEM). The protons are attracted to the cathode by the oxygen that is flowing to the cathode from the flow field plate. While the protons move through the PEM, the electrons cannot pass

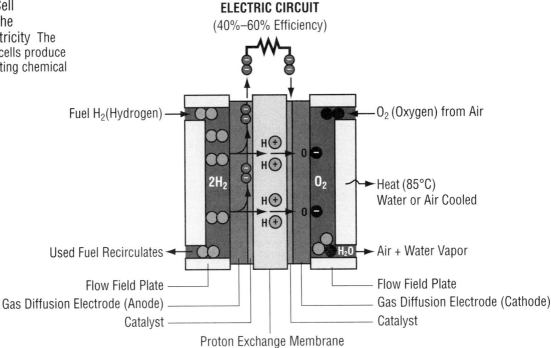

Figure 3-9 Fuel Cell Components and the Generation of Electricity The components in fuel cells produce electricity by converting chemical energy.

ELECTRIC CIRCUIT
(40%–60% Efficiency)

Fuel H$_2$(Hydrogen)

O$_2$ (Oxygen) from Air

2H$_2$

O$_2$

Heat (85°C)
Water or Air Cooled

Used Fuel Recirculates

H$_2$O → Air + Water Vapor

Flow Field Plate
Gas Diffusion Electrode (Anode)
Catalyst
Proton Exchange Membrane
Flow Field Plate
Gas Diffusion Electrode (Cathode)
Catalyst

(Courtesy of Ballard Power Systems)

through the PEM and must move through an external circuit. As the electrons pass through the external circuit, they are used as a source of electricity. After they pass through the external circuit over to the cathode side, they then combine with the protons, oxygen, and catalyst to create water vapor, which is emitted. Because water vapor is the primary by-product (exhaust), fuel cells are considered to be a cleaner form of energy that causes less pollution.

3. Safety Benefits and Risks Associated with Fuel Cells

Energy sources such as gasoline and natural gas are known to be volatile, and taking precautions to minimize the chance of an explosion is always of concern. Hydrogen is the fuel used in fuel cell technology, and it too has the potential of exploding. However, it does not have the same volatile potential as other fuels. Gasoline becomes volatile when it reaches a concentration level of only 1 percent. Hydrogen requires a 4 percent concentration level before it becomes volatile.

It would seem that since hydrogen has to be four times more concentrated than gasoline before it becomes volatile, then it would be safer. But is it? It depends on what concentrations of hydrogen are needed in the onboard storage tanks of fuel cell cars. If hydrogen is concentrated in high enough amounts, then it is explosive, just like gasoline is in the cars on the road today.

The volatility of hydrogen versus traditional fuels is only one area in which to assess the benefits and risks. Other issues of comparison include:

- Toxicity in the event of a spill
- Amount of pollution from emissions
- Cost of obtaining and storing hydrogen

4. The Hydrogen Economy

Fuel cell technology and the hydrogen economy are rapidly growing and are touted as being the technology that will transform our economy and society. U.S. dependency on importing foreign oil is well known. There has always been concern that when relations with foreign countries change, so does the availability and cost of importing oil. Of more concern recently is the diminishing supply of fossil fuels. Many countries consume much more energy than they produce. Since fossil fuels are not renewable sources of energy but instead must be obtained from the earth, some people, including geologists, are concerned that if the rate of consumption continues to grow, the demand for global oil reserves will exceed supply.

This possibility would have devastating effects. If oil consumption peaked and reserves began to become depleted, using oil as the source of energy would become virtually unaffordable because the demand would be so much greater than the supply. This means that all economies would be affected since most goods and services rely on transportation, which relies on the oil reserves. For this reason, developing technology that uses an energy source that is renewable and abundant is of primary importance.

Hydrogen may well be the answer because it is a colorless, odorless, non-toxic gas. Hydrogen in pure form is not found in nature by digging or mining like coal. It typically exists in the form of chemical bonds such as those found in water: H_2O. Hydrogen is abundant in various compound forms and must be processed to be extracted for use. The process of getting pure hydrogen does not require using energy from sources such as fossil fuels, but can instead be obtained using solar, water, or **wind energy**. The process of creating hydrogen is not new. In fact, approximately 500 billion cubic meters of hydrogen are produced, stored, and transported worldwide every year.

Hydrogen is created from many sources. Steam reforming and partial oxidation are two methods of creating hydrogen from fossil fuels. Electrolysis is the process used to create hydrogen from water. There are different forms of electrolysis, including conventional, high-pressure, and high-temperature electrolysis. Hydrogen can also be generated from **biomass**. Technologies that could be used commercially to generate hydrogen from biomass are currently being researched and developed.

5. Putting It All Together

This project has introduced the topic of fuel cell technology as an example of an alternative energy source. Fuel cells use hydrogen as an energy carrier. Through a chemical process that occurs in the proton exchange membrane (PEM), hydrogen electrons are used as a source of electricity.

Using hydrogen as an energy source may have profound implications on sustaining the current economy as well as global resources that are currently being consumed at an alarmingly high rate compared to production.

The design brief activity for this project will provide an opportunity to synthesize the information from this reading with additional information found through researching current literature. You will develop a comprehensive presentation of fuel cell technology and the hydrogen economy.

Key Terms

energy	geothermal	membrane	electrode
electricity	solar energy	electrode	anode
fossil fuels	hydrogen energy	assembly (MEA)	cathode
hydroelectric	fuel cell car	proton exchange	wind energy
wind	fuel cell	membrane	biomass
nuclear	chemical energy	(PEM)	

III. Student Materials

Use Project 5 Design Brief: Fuel Cell Technology—An Overview to complete the project.

UNIT 3: PROJECT 5   DESIGN BRIEF: FUEL CELL TECHNOLOGY—AN OVERVIEW

You are a representative from Global Economy Solutions and your job is to present to the U.S. Department of Energy a report on fuel cell technology. This report will be used to determine how much funding should be given to researchers who are studying and developing fuel cell technologies.

The report must be comprehensive and must include:

- Why fuel cell technology is needed

- How it works

- Its benefits and risks compared to fossil fuels

- Its current status

  ○ How is it currently being used?

  ○ Are products that use fuel cell technology currently commercially available?

- A graphic of a commercially available product or use of fuel cell technology

- Where to find out more information about fuel cell technology

Design Constraints (Things You Must Do)

- Research fossil fuels and include some basic information on why we should research new sources of energy (e.g., discuss the limitations, risks, and costs).

- Create a graphic of an individual fuel cell. Make sure to include and label the components (i.e., flow field plates, catalyst, anode, cathode, and PEM).

- Based on your graphic, explain how an individual fuel cell works.

- Provide a brief comparison of fuel cells versus fossil fuel on topics such as safety, pollution, and toxicity.

- Provide three sources of additional information regarding fuel cell technology.

  - One source must be the URL of a government source on fuel cell technology or initiatives.

  - One source must be for a commercial product or use of fuel cell technology.

  - The third source is your choice.

PROJECT 6
Small-scale Hydroelectric Power

I. Project Lesson Plan

1. Project Description

You will explore hydroelectric power as a renewable energy source. In particular, you will develop an understanding of how the potential and kinetic energy of water can be converted into work. The project includes a brief history and description of hydroelectric power. You will examine and create visualizations of the main components of a typical hydroelectric system. This project culminates in the *Run-of-the-River* activity in which you conduct and report upon a Preliminary Feasibility Assessment of a potential small-scale or microhydropower source. You will also be asked to consider local requirements and restrictions, as well as the ecological and economical impact of building the system.

2. Learning Objectives

- You will gain an appreciation for the history and current state of hydropower usage.
- You will describe how energy can be harnessed from moving water and identify the major components of a hydropower system.
- You will detail the common classifications for hydropower systems.
- You will examine the efficacy of creating a local microscale hydroelectric system.
- You will assess the economic and ecological impact of creating a small-scale hydropower system.

II. Background Information

1. A Brief History of Hydroelectric Power

Hydroelectric power systems have a long and interesting history; in fact, people have been harnessing water to perform work for thousands of years. More than 2000 years ago the Greeks used water wheels to power enormous grinding stones that ground wheat into flour. The 1700s and 1800s saw numerous developments that ultimately led to the modern water **turbine** and today's hydropower systems. For example, in 1880 a brush arc light **dynamo** driven by a water turbine was used to provide theatre and storefront lighting in Grand Rapids, Michigan. In 1881, a similar brush **dynamo** connected to a turbine in a flour mill provided street lighting at Niagara Falls, New York. However, these two projects used **direct-current (DC)** technology and it wasn't until 1887 and Nikola Tesla's invention of a generator that produced **alternating-current (AC)**, that hydropower systems were able to produce and transmit electricity over long distances. With Tesla's invention came the relatively rapid and widespread use of hydroelectric power. See the **Resources Index** for Web sites with more information on the history of hydroelectric power and the work of Nikola Tesla.

2. Usage Statistics

Hydropower is the most often used type of renewable energy source. It accounts for more than 90 percent of all electricity that comes from renewable resources (i.e., solar, geothermal, wind, and biomass). Worldwide, about 20 percent of all electricity is generated through the use of water. Canada leads the world in hydroelectric projects and output, while Norway produces more than 99 percent of its electricity with hydropower.

In the United States, hydropower accounts for up to 10 percent of the nation's supply of electricity. That is more than 95,000 **megawatts (MW)** of electricity annually, which is enough to meet the needs of about 35 million residential customers in California, New York, Ohio, Pennsylvania, and Texas. See the **Resources Index** for Web sites with more usage statistics.

In fact, the United States is the second largest producer of hydropower in the world. However, over half (56 %) of the total U.S. hydroelectric capacity for electricity generation is concentrated in three states: Washington, California, and Oregon (Figure 3-10). Approximately 31 percent of our nation's hydropower generation is in Washington, the location of the nation's largest hydroelectric facility, the Grand Coulee Dam. This may be in part due to the fact that hydropower is generated at only 3 percent of the nation's 80,000 dams.

3. Classes of Hydroelectric Power Systems

Hydropower plants range in size from small systems for a home or village to large projects producing electricity for utilities. The vast majority of the hydropower produced in the United States comes from *large-scale* projects that generate more than 30 megawatts (MW). This is enough electricity to power nearly 30,000 households (Table 3-3). *Small-scale hydropower systems* are those that generate between .01 to 30 MW of electricity. Hydropower systems that generate up to 100 **kilowatts** (kW) of electricity are often called *microhydro systems.* Most of the systems used by home and small business owners would qualify as microhydro systems. In fact, a 10 kW or *Picohydro* system can provide enough power for a large home (see the Resource Index under kilowatts and megawatts for more information).

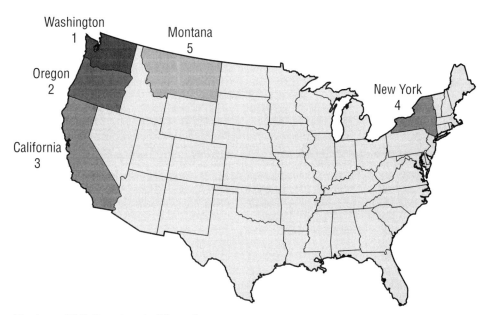

Figure 3-10 Distribution of Hydropower Producing States A map of the leading hydropower producing states.

(Courtesy of U.S. Department of Energy)

Table 3-3 Classes and Sizes of Hydropower Systems

The names and corresponding size classifications of hydroelectric power systems are listed.

Type of Hydropower System	Energy Output
Large-scale	30 MW (30,000 kW) or more
Small-scale	0.01 MW (10 kW) to 30 MW (30,000 kW)
Microhydro	0.01 MW (10 kW) to 0.1 MW (100 kW)
Picohydro	Up to 0.01 MW (10 kW)

4. How Hydroelectric Power Works

In the simplest sense, **hydropower** refers to the use of water to power machinery and/or make electricity. The basic principle behind hydropower involves capitalizing on the **kinetic energy** of flowing water as it moves downstream and converting it into mechanical energy. If water can be directed, harnessed, and/or channeled from a higher level to a lower level, the resulting water pressure can be used to do *work*. **Turbines** convert water pressure into mechanical shaft power, which can be used to drive an electricity **generator**. In turn, this generator converts the energy into electricity and this electricity is then fed into an **electrical grid** to be used in homes, businesses, and by industry. Because the water cycle (Figure 3-11) is an endless, constantly recharging system, hydropower is considered a **renewable energy**. Solar energy does the work of raising the water to a higher level. Water is evaporated by the sun, rises, and forms clouds. Water that falls at higher elevations flows into streams and rivers, where it can be harnessed for hydropower.

The amount of available energy in moving water is determined by its **flow** or fall. Swiftly flowing water in a big river, such as the Columbia River (along the

Figure 3-11 The Water Cycle An illustration of the water cycle upon which hydropower ultimately relies.

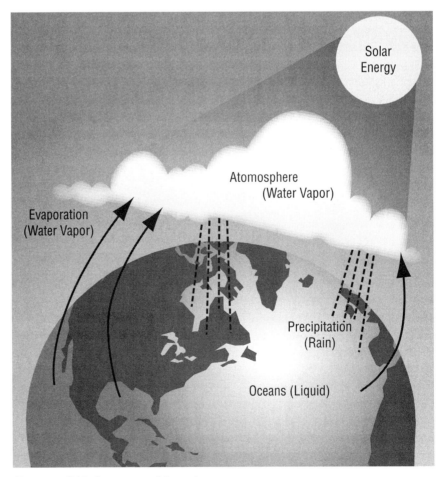

Solar Energy

Atomosphere (Water Vapor)

Evaporation (Water Vapor)

Precipitation (Rain)

Oceans (Liquid)

(Courtesy of U.S. Department of Energy)

border of Oregon and Washington), carries a great deal of energy in its flow. So, too, does water descending rapidly from a very high point, such as Niagara Falls in New York. In either case, the water flows through a pipe, or **penstock**, then pushes against and turns blades in a **turbine** to spin a **generator** to produce electricity.

There are three basic types of hydropower facilities or systems: **impoundment, diversion**, and **pumped storage**. The most common type of hydroelectric power plant is an impoundment or storage facility. These are typically large-scale hydropower systems that use a dams to store river water in reservoirs. Water released from the reservoir flows through a penstock into a turbine. The water spins the turbine, which in turn activates a generator to produce electricity. See Figure 3-12 for a diagram of a typical impoundment system. The water flow may be regulated to meet fluctuating electricity needs or to maintain a constant level in the reservoir. The second type of system, a pumped storage system, is similar to an impoundment system. However, when the demand for electricity is low, a pumped storage facility stores energy by pumping water from a lower reservoir to an upper reservoir. During periods of high electrical demand, the water is released back to the lower reservoir to generate electricity.

The last type of hydroelectric or hydropower system is a diversion or **Run-of-the-River system**. These systems are generally used in micro or picohydro facilities. Here a portion of a river or stream's water is diverted to a channel, pipeline, or penstock that delivers power directly to the waterwheel or turbine. The force of the current applies the needed pressure, and these projects do not require dams or large storage reservoirs. Figure 3-13 shows the parts and layout of this type of hydropower project.

In these systems, the water is first funneled through a series of structures that control its flow and filter out debris. A *headrace* is sometimes necessary if there is an insufficient head. This headrace is a cement or masonry waterway running parallel to the water source and leads to the *forebay* (also concrete or masonry). Its role is to act as a settling pond for large debris that might otherwise flow into the system and damage the turbine. Additional filtration is sometimes provided by a *trashrack* that, as its name suggests, is a grill that removes additional trash and debris. Once filtered, the water flows into a penstock, which directs it to the turbine and generator found in the *powerhouse*. These conveyances can be constructed from plastic pipe, cement, steel, and even wood.

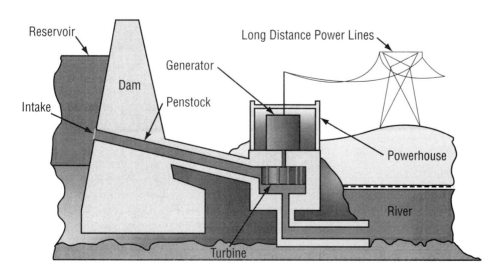

Figure 3-12 Representative Illustration of a Hydropower System The main components of a storage type hydroelectric system.

(Courtesy of PPL Corporation)

Figure 3-13 Parts and Layout of a Diversion or Run-of-the-River Hydropower System

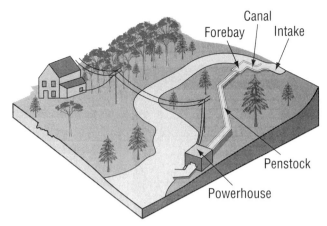

(Courtesy of U.S. Department of Energy)

5. Feasibility Assessment of a Microhydroelectric System

One can conduct a *Preliminary Feasibility Assessment* of a potential small-scale or microhydropower project relatively easily by following the steps outlined in the following text. The first step is to identify and find flowing water. This can be a stream or small river that has a sufficient quantity of falling water. If a potential source is not known, the *U.S. Geological Survey* (USGS) maintains *The National Streamflow Information Program*. Here you can access interactive *Streamgage* maps and data for your state. See **Resource Index** under Locating a Source for more information. Next, you'll want to determine the amount of power that you can obtain from the flowing water on your site. The power available at any instant is the product of the **head** and **flow** of your water source.

a. Determining the Head

Your source's head is essentially the vertical distance that water falls (Figure 3-14), measured in feet, meters, or units of pressure; it is also influenced by the type of water conveyance that is employed.

Be sure to locate a source of water that has an adequate head because, as you might suspect, the higher the head the less water you will need to produce a given amount of power. Additionally, with a high head flow, you can use smaller, less expensive equipment. If the head is too low (a change in elevation of less than 2 feet or 0.6 meters), a small-scale hydroelectric system may be unfeasible. To determine the head, you need to consider both *gross head* (the vertical distance between the top of the penstock and the point where

Figure 3-14 An Illustration of a Run-of-the-River Project's Head

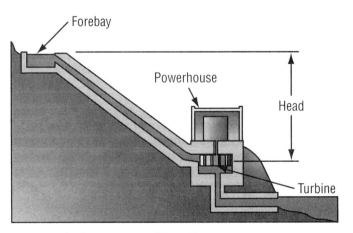

(Courtesy of U.S. Department of Energy)

the water discharges from the turbine) and *net head*. The net head equals gross head minus losses due to friction and turbulence in the piping.

You can use several different methods to get an estimate of the gross head's vertical distance. You can use U.S. Geological Survey maps of your area (for more information on these maps, see the **Resource Index** under Maps). Another option is to gather data about the elevation changes and gross head of a potential source by using a hand-held Global Positioning System (GPS) device. For more information, see the **Resource Index** under GPS Devices. The *hose-tube* method can also be used to determine the gross head of your stream of river. The **Preliminary Feasibility Assessment Guidelines** provide a detailed explanation of this technique.

b. Determining the Flow

The flow or quantity of water falling is the volume of water passing a point in a given amount of time and is measured in gallons per minute, cubic feet per second, or cubic meters per second. For very small streams, you can measure the flow of your source directly using the *bucket method*. This involves damming your stream to divert its flow into a bucket or container. The rate at which the container fills is the flow rate. For example, a 5-gallon bucket that fills in 5 minutes indicates that your stream's water is flowing at 1 gallon per minute (gpm). Note that 1 cubic meter per second (m^3/s) equals 15,842 gpm. So in our example, the flow rate of 1 gpm is equivalent to 0.0000631 m^3/s.

For larger streams and rivers, whose water flow is not easily diverted, you could use the *float method* to measure its flow rate. Start by creating a cross-sectional profile of your streambed and establishing an average cross-section for a known length of stream. To do this, select a stretch of the stream with the straightest channel and most uniform depth and width as possible. At the narrowest point of this stretch, measure the width of the stream by using a tape measure. Then, with the yardstick or your calibrated measuring rod, walk across the stream and measure the water depth at 1-foot (or 30-centimeters) increments across the entire stream (Figure 3-15). The cross-sectional area of the stream is determined by multiplying channel depth (D) by channel width (W) along this transverse section of the stream. For a hypothetical stream with a rectangular cross-sectional shape (a stream with a flat bottom and vertical sides), the cross-sectional area (A) is simply the width (W) multiplied by the depth (D) or $A = W \times D$.

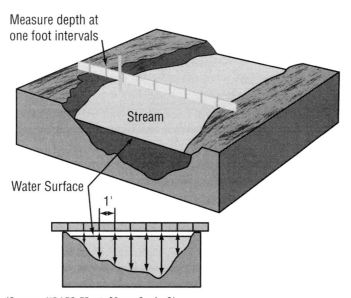

Figure 3-15 Illustration of How to Determine the Cross-sectional Area of a Stream

(Source: NRAES FS-13 [Out of print])

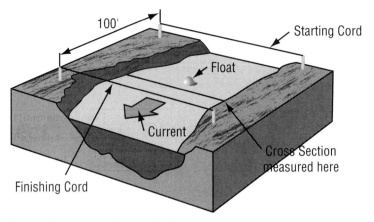

Figure 3-16 An Illustration of How to Determine the Flow Velocity of a Stream

100'

Starting Cord

Float

Current

Cross Section measured here

Finishing Cord

(Source: NRAES FS-13 [Out of Print])

Next, use a float (e.g., ping-pong ball, fishing float, or pieces of wood) and time how long it takes to travel a measured length of stream (Figure 3-16). A 50-foot (15.24 meters) to 100-foot (30.48 meters) stretch will suffice. Conduct several trials and calculate the average time it takes to travel this distance. Divide this average time by distance to obtain an average or mean flow velocity.

The principle of velocity-area methods suggests that flow (Q) equals the mean velocity (V_{mean}) multiplied by the cross-sectional area (A):

$$Q \text{ (m}^3\text{/s)} = V_{mean} \text{ (m/s)} \times A \text{ (m}^2\text{)}$$

This velocity must be reduced by a correction factor, which estimates the mean velocity as opposed to the surface velocity. By multiplying averaged and corrected flow velocity, the volume flow rate is estimated. This method provides only an approximate estimate of the flow. Approximate correction factors to convert measured surface velocity to mean velocity are as follows:

Concrete channel, rectangular, smooth	0.85
Large, slow, clear stream	0.75
Small, slow, clear stream	0.65
Shallow (less than 0.5 m / 1.5 ft.) turbulent stream	0.45
Very shallow, rocky stream	0.25

The **Resource Index** contains more conversion factors that may be needed as you assess your site's feasibility. A less engaging way to determine the flow is to consult the USGS's National Streamflow Information Program and its Streamgage maps mentioned earlier.

c. Estimating Power Output

Once the head and flow of your water source is known, these figures can be used to estimate the power output for the system. The power generated is represented by the equation:

$$P = e \times H \times Q \times g$$

where:

P = electric power output in kilowatts (kW)
e = efficiency range 0.75 to 0.88 (75% to 88%)
H = head, in meters (m)
Q = flow, in cubic meters/sec (m³/s)
g = acceleration of gravity, normally 9.81 m/s/s

Efficiency is a percentage obtained by dividing the actual power or energy by the theoretical power or energy. It represents how well the hydropower plant converts the energy of the water into electrical energy. For small-scale hydroelectric applications, if an efficiency value of 81 percent is assumed, the following equation can be used:

$$P \text{ (kW)} = 0.81 \times H \text{ (m)} \times Q \text{ (m}^3\text{/s)}$$

6. Equipment Costs and Availability

Few companies make very small or microhydropower turbines. Low head, low flow turbines may be difficult to find. Commercially available turbines and generators are usually sold as a package. You may be able to find and refurbish old but operable turbines but these "do-it-yourself" systems require careful matching of a generator with the turbine horsepower and speed. See the **Resource Index** under Equipment Costs and Availability for more information.

7. Requirements and Restrictions

It should be noted that a water flow's use, access to, control, or diversion of it is regulated to a relatively high degree in this country. There are many local, state, and federal regulations that govern the construction and operation of a hydroelectric system. However, if the project presents a minimal physical impact (as in the case of most run-of-the-river systems), the legal process may not be terribly complex. It also helps if you do not plan to sell the energy produced.

When planning the development of a system, your initial point of contact should be the county engineer, for he or she will most likely be well informed about the restrictions in your particular area. Additionally, your state energy office may be able to provide you with insight into the process. You'll also need to determine how much water you can divert from your stream channel. Each state controls its own water rights. You can find information about your state's energy office through the *National Association of State Energy Officials* (NASEO), which provides current contact information for state energy offices, including links to their Web sites.

It may also be prudent to contact the *Federal Energy Regulatory Commission* (FERC) and/or the *U.S. Army Corps of Engineers*. These are the main federal agencies that will influence the planning and development of your system. For more information, see the **Resource Index** under Local Requirements and Restrictions.

8. Risks and Benefits of Hydroelectric Power

The potential benefits of micro and small hydroelectric power systems makes them attractive options for meeting part of the future energy demands, particularly in rural areas. First the financial investment required is lower than that for conventional, centralized energy systems. Moreover, hydropower is an extremely clean and renewable source of energy. In fact, it uses only the water (no water is actually consumed), so the water is available for other purposes such as irrigation, recreation, and water supply. Also, no fuel and limited maintenance are required, so running costs are relatively low in comparison to diesel power. Lastly, it is a long-lasting and robust technology. Well-designed and well-built systems can last for 50 years or more without major new investments.

But hydropower is not without risks and shortcomings. The main concerns with this technology are ecological (i.e., water quality and habitat condition). Many systems create barriers to upstream and downstream fish passage, and fish populations can be negatively impacted if fish cannot migrate upstream past impoundment dams to spawning grounds or if they cannot migrate downstream to the ocean. Hydropower may also have deleterious effects on water quality and flow. Hydropower plants can cause low dissolved oxygen levels in the water, a problem that can be harmful to the flora and fauna in the stream/riverbank habitats. Finally, hydropower is inextricably linked to the water cycle and in times of drought when water is not available, hydropower plants cannot produce electricity.

The Canadian Renewable Energy Network (CanREN) outlines how building a hydroelectric system impacts the environment. It also describes "preventable measures" that can be taken to lessen or avoid any negative effects of this technology. To view this list, see the **Resource Index** under Risks and Benefits of Hydroelectric Power.

There also exists the possibility of using software to conduct a preliminary feasibility assessment. The *Energy Diversification Research Laboratory (CEDRL)* has developed free downloadable software called *RETScreen*. It allows you to perform a preliminary site evaluation study of potential small-scale hydroelectric sites. For more information, see the **Resource Index** under Preliminary Feasibility Assessment Software.

Key Terms

hydroelectric power	megawatt	renewable energy	pumped storage (system)
turbine	kilowatt	flow	run-of-the-river system
dynamo	hydropower	penstock	
direct-current (DC)	kinetic energy	impoundment (system)	
alternating-current (AC)	generator	diversion (system)	
	electrical grid		

III. Extended Background Materials

More About Hydropower Turbines and Generators

The type of hydropower turbine used in a project often depends on several factors including the height of standing water (referred to as the head), the flow (volume) of the water, efficiency, and cost. There are two basic types of hydropower turbines: *impulse* and *reaction*. An impulse turbine uses the velocity and force of the flowing water to move a *runner* (the rotating part of the turbine). The water stream hits *buckets* or blades on the runner and turns it as depicted in Figure 3-17.

There is no suction on the down side of the turbine, and the water flows out the bottom of the turbine housing after hitting the runner. An impulse turbine is generally suitable for high head, low flow applications. An example of an impulse turbine is the Pelton turbine.

Figure 3-17 Illustration of an Impulse Turbine

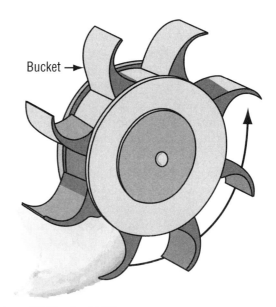

Bucket →

(Courtesy of Wikipedia, 2005)

A reaction turbine generates power by combining pressure and moving water. The runner is placed directly in the water stream flowing over the blades rather than striking each individually. Reaction turbines, rather than impulse turbines, are generally used for sites with lower head and higher flows. For more information on the different types of turbines, see the **Resources Index**.

Another key component to any hydropower system is the generator. Figure 3-18 shows a cutaway of a typical generator, and its basic described in the following:

1. As the water turns the turbine, the turbine turns a shaft that rotates a series of magnets past copper coils and a generator to produce electricity.

2. The *Kaplan Head* is associated with adjustable blades on the turbine. Adjustable blades operate efficiently despite variations in water flow and energy demands.

3. The *rotor* is the rotating part of a generator and is essentially a series of magnets that generate a magnetic field as the spin.

4. The *stator* is the stationary part of the generator made of coils of copper wire and electricity is produced as the rotors spin past the stationary wiring.

5. A shaft connects the turbine to the rotor, and all three components (the turbine, shaft, and rotor) turn at the same speed.

6. A series of adjustable vanes called wicket gates control the volume of water flowing through the turbine.

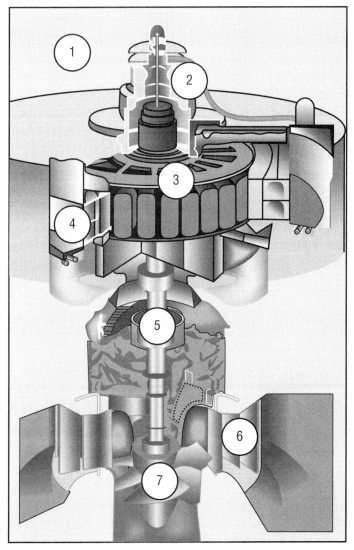

Figure 3-18 A Cutaway of a Typical Hydropower Generator

(Courtesy of The Foundation for Water and Energy Education)

Once the AC electricity is created, transformers increase the voltage of the electricity produced by the generator to levels necessary for its transport to our communities via transmission lines. Higher voltages move electricity over long distances most efficiently. As the electricity is transported to local power lines and eventually individual homes and businesses, the voltage is stepped down.

IV. Student Materials

Use Project 6 Design Brief: The *Run-of-the-River* Project and Preliminary Feasibility Assessment Guidelines to complete the project.

UNIT 3: PROJECT 6   DESIGN BRIEF: THE "RUN-OF-THE-RIVER" PROJECT

As a recent college graduate with a BS degree in environmental management and sustainability, you have recently accepted a position as a consultant for the National Hydropower Association. Your primary responsibility is to increase the general public's awareness of the potential of hydroelectric power in the United States. You learned in your classes that hydropower is one of the nation's most important renewable energy resources and the technology for producing hydroelectricity has significant advantages over other energy sources. You recognize that it is a reliable, domestic, renewable resource with large undeveloped potential.

But you also realize that the current capacity of hydropower in the United States is about 80,000 megawatts (MW), which is produced at about 180 federal projects and more than 2000 non-federal projects. This represents the power generated from only 3 percent of the nation's 80,000 dams, and the nation's generation is predicted to decline through 2020 (due, in part, to environmental issues, regulatory complexity, and changes in energy economics). Only 560 MW of conventional hydropower capacity is expected to be added by 2025. With this looming decline in the development and utilization of hydropower in mind, you have been given the task of informing the public and gaining support for this technology.

Your first assignment is to travel the rural areas of your region and visit local farmers. Your boss said she wants you to "work with them to be sure they are aware of hydropower's great potential and that they are able to intelligently weigh the costs and benefits of the technology." Before she sends you on the road, she wants to be sure you "know your stuff" and expects you to conduct and report upon a Preliminary Feasibility Assessment of a potential small-scale or microhydropower source.

Design Constraints (Things You Must Do)

The report must be comprehensive and must include:

- A brief introduction to the hydroelectric or hydropower technology and its current status.
- A detailed and scientifically accurate description of how the technology works.
- The results of your preliminary feasibility assessment on a local stream, including:
    - Location and description of your source
    - Calculations and results of your head, flow, and power assessments
    - Sketch or plan of what your system will look like including its necessary parts or components
    - Budget outlining the cost and availability of the parts and equipment necessary for the deigned system
    - Summary of the local requirements and restrictions

- Brief ecological and economical cost/benefit analysis
- Where to find out more information about small-scale hydroelectric technology
  - URL of a government source on hydropower technology or initiatives
  - Commercial product or use of hydropower technology
  - Source of your choice

PRELIMINARY FEASIBILITY ASSESSMENT GUIDELINES

A small-scale or micro hydroelectric facility requires an adequate head of water and a sizable flow of water. One can conduct a Preliminary Feasibility Assessment of a potential small-scale or microhydro power project relatively easily by following the steps outlined here.

1. Identify and find flowing water. This can be a stream or small river that has a sufficient quantity of falling water. Consult the *U.S. Geological Survey's* (USGS) *Streamflow Information Program,* where you can access interactive *Streamgage* maps and data for your area.

2. Once a potential source is located, you will need to determine the amount of power that you can obtain from the flowing water on your site. The power available at any instant is the product of the **head** and **flow** of your water source.

Determining the Head

Your source's head is the vertical distance that water falls—measured in feet, meters, or units of pressure—and is illustrated in Figure 3-14 on page 119.

You can estimate the head's vertical distance several ways:

1. Use U.S. Geological Survey maps of your area.
2. Use a hand-held Global Positioning System (GPS) device.
3. The *hose-tube method* is detailed in the following section.

Hose-tube Method

The *hose-tube method* involves taking stream-depth measurements across the width of the stream you intend to use for your system—from the point at which you want to place the penstock to the point at which you want to place the turbine.

You will need an assistant and:

- 20- to 30-foot (6 to 9 meters) length of small-diameter garden hose or other flexible tubing
- Funnel
- Yardstick or measuring tape
- Calculator

Stretch the hose or tubing down the stream channel from the point that is the most practical elevation for the penstock intake.

1. Have your assistant hold the upstream end of the hose, with the funnel in it, underwater as near the surface as possible.

2. Lift the downstream end until water stops flowing from it.

3. Measure the vertical distance between your end of the tube and the surface of the water. This is the gross head for that section of stream.

4. Have your assistant move to where you are and place the funnel at the same point where you took your measurement. Then walk downstream and repeat the procedure.

5. Continue taking measurements until you reach the point where you plan to site the turbine.

6. The sum of these measurements will give you a rough approximation of the gross head for your site.

Note: Due to the water's force into the upstream end of the hose, water may continue to move through the hose after both ends of the hose are actually level. You may wish to subtract an inch or two (2 to 5 centimeters) from each measurement to account for this. It is best to be conservative in these preliminary head measurements.

Determining the Flow

The **flow** or quantity of water falling is measured in gallons per minute, cubic feet per second, or liters per second. You can measure the flow of your source directly using the *bucket method.*

Bucket Method

You will need an assistant and:

- Large bucket (5 gallons or greater) or other large container
- Stopwatch
- Calculator

Dam your stream to divert its flow into a bucket or other large container.

1. Time how long it takes for the bucket to fill up. You may want to do this multiple times at various locations along the water flow and generate an average measure.

2. The rate at which the container fills is the flow rate for your source. For example, if a 5-gallon bucket fills in 5 minutes, this indicates that your stream's water is flowing at 1 gallon per minute (gpm).

Note: 1 cubic meter per second (m^3/s) = 15,842 gpm. So, in our example, the flow rate of 1 gpm is equivalent to 0.0000631 m^3/s.

For larger streams and rivers, where water flow is not easily diverted, you could use the *float method* to measure the flow rate.

Float Method

You will need an assistant and:

- Tape measure
- Meter stick or a calibrated measuring rod
- Stopwatch
- Floats

1. Create a cross-sectional profile of your streambed by establishing an average cross-section for a known length of stream. To do this, select a stretch of the stream with the straightest channel and most uniform depth and width as possible.

2. At the narrowest point of this stretch, measure the width of the stream by using a tape measure.

3. Then, with the yardstick or your calibrated measuring rod, walk across the stream and measure the water depth at 1-foot (or 30 centimeters) increments across the entire stream.

4. Calculate the cross-sectional area of the stream by multiplying channel depth (D) by channel width (W) along this transverse section of the stream ($A = W \times D$).

5. Use a float (e.g., a ping-pong ball, a fishing float, or pieces of wood) and time how long it takes to travel a measured length of stream. A 50-foot (15.24 meters) to 100-foot (30.48 meters) stretch will suffice.

6. Conduct several trials and average the time it takes the float to travel this distance. Divide this average time by distance to obtain an average or mean flow velocity.

7. Calculate your stream's flow rate using the following equation:

$$Q \ (\text{m}^3/\text{s}) = V_{mean} \ (\text{m/s}) \times A \ (\text{m}^2)$$

Where flow (Q) equals the mean velocity (V_{mean}) multiplied by the cross-sectional area (A).

This velocity must be reduced by a correction factor, which estimates the mean velocity as opposed to the surface velocity. By multiplying averaged and corrected flow velocity, the volume flow rate is estimated. Approximate correction factors to convert measured surface velocity to mean velocity are as follows:

- Concrete channel, rectangular, smooth   0.85
- Large, slow, clear stream   0.75
- Small, slow, clear stream   0.65
- Shallow (less than 0.5 m / 1.5 ft.) turbulent stream   0.45
- Very shallow, rocky stream   0.25

See Figure 3-15 and Figure 3-16 for guidance.

Remember to consider the state and federal requirements and restrictions that may apply to this site!

Don't forget your budget!

UNIT 4
Nanotechnology

Unit Overview

I. Introduction

The term *nanotechnology* is quickly becoming a familiar term. But what is nanotechnology and what does it have to do with education? *Nanotechnology* is a term often used in a very general manner to refer to all aspects of nano. Nano, however, is truly a multidisciplinary field with individuals from fields such as chemistry, physics, biology, materials science, and engineering. All of these individuals are working to better understand and apply knowledge of objects that are nanoscale in size. The nanoscale dimension is a range from hundreds to tens of nanometers (0.1 to 100 nm) or 10^{-9}. To give some perspective of just how small a nanometer is, one meter equals one billion nanometers. A human hair is roughly 50,000 nanometers across the tip.

To simplify, the individuals working in the field of nano can be categorized as either nanoscientists or nanotechnologists depending on what they are trying to do. Nanoscientists try to figure out the fundamental physical, chemical, and biological properties of objects (e.g., atoms, molecules, and structures) at the nanoscale. Nanotechnology is basically the application of nanoscience. It consists of the tools and techniques used to research, develop, and produce nanoscale objects and devices.

According to the chair of the Subcommittee on Nanoscale Science, Engineering and Technology, 2 million nanotech-trained workers will be needed in the next 10 to 15 years. To be responsible consumers of the information regarding nanotechnology and to have a workforce prepared for the jobs of tomorrow, it is important to learn about the field of nanotechnology.

II. Unit Learning Goals

- You will learn the key terminology and concepts associated with the field of nano.
- You will develop some perspective for how small the nanoscale dimension is and the constraints of working at this small scale.
- You will learn about atomic and molecular level forces and how they affect the properties and characteristics of materials.
- You will research and evaluate current information regarding nanoscience and nanotechnology to prepare a comprehensive overview presentation that concludes with why it is important to know about this emerging field.
- You will gain an understanding of how the scanning tunneling microscope and the atomic force microscope enable scientists to work at the nanoscale.
- You will explore the societal issues raised by advances in nanoscience and nanotechnology.

III. Projects

Introductory level

PROJECT 1: Overview of the Field and the Nanoscale

During this introductory project, you will become familiar with the key terms, definitions, and distinctions that are part of the inherently multidisciplinary field of nano. A focus will be placed on developing your understandings of what a nanometer is, as well as the relative and absolute size of things. This project also engages you in activities in which you use your senses to conceptualize the nanoscale.

PROJECT 2: What Is Nanoscience?

During this project, you will explore the multidisciplinary field of nanoscience more deeply and learn how nanoscientists plan to build with atoms. In doing so, you will be introduced to the forces and effects that take place at the nanoscale. Additionally, you will discover the way in which the characteristics and properties of matter differ when one moves from the macroscale to the nanoscale.

PROJECT 3: Tools of Nanotechnology

You will learn about the tools that enable nanotechnologists to research, design, and nanoengineer at the nanoscale. This project begins with a description of the relationship between nanoscience and nanotechnology. Next, you are introduced to the common tools and techniques that scientists in the field of nanotechnology use in their work. You will use current technology to generate an image by using a procedure similar to that of scanning probe microscopy.

PROJECT 4: Ethical and Societal Concerns Related to Nanotechnology

In this project, you will examine and report on the potential risks and benefits of advances in the field of nanotechnology. Nanotechnology has been promoted as the next technological revolution and it has the potential to produce new materials that are substantially stronger while weighing less. It also has the potential to change both medicine and defense. But do breakthroughs in nanotechnology present unique health and environmental dangers that need to be studied? As with any new technology, there are several concerns regarding its impact in the areas of health, environment, privacy, the workforce, and the economy. This project culminates in a design brief in which you research the risks and benefits of this emerging technology. You design, conduct, and report upon a survey aimed at assessing the general public's level of understanding and opinions regarding nanotechnology.

Intermediate Level

PROJECT 5: Forces and Effects at the Nanoscale

During this project, you will explore the way in which the characteristics and properties of matter differ when one moves from the macroscale to the nanoscale. You will actively investigate several of the key forces and effects (e.g., surface effects and self-assembly) at this small scale.

PROJECT 6: Applications in the Field of Nanotechnology

In this project, you will extend your knowledge of cutting edge applications of nanotechnology. Using the background on nanotechnology provided here and in Project 3, you will examine some of the current and future applications of nanotechnology in the areas of manufacturing, medicine, and defense. The project culminates in a design brief in which you assume the role of a design engineer for a private nanotechnology firm whose task is the development of a presentation at a local town meeting.

Advanced Project

You will complete an independent project through the use of visualization tools by researching a new topic dealing with bioprocessing or by expanding on topics covered in this unit. The objective of the advanced level projects is for you to further your skills in integrating research, problem solving through the design brief approach, and presentation. It is up to the teacher to work with you to negotiate the topic, time allocated to the project, and design constraints.

IV. Unit Resources

The Resource Index contains a listing of all resources associated with the unit. Included are relevant Web sites, books, and other publications. The Glossary provides definitions for all Key Terms listed in each project.

PROJECT 1
Overview of the Field and the Nanoscale

I. Project Lesson Plan

1. Project Description

During this introductory project, you will become familiar with the key terms, definitions, and distinctions that are part of the inherently multidisciplinary field of nano. You will increase your understanding of what a nanometer is, as well as the relative and absolute size of things. This project also engages you in activities in which you will use your senses to conceptualize the nanoscale.

2. Learning Objectives

- You should be able to define the key terminology and explain the major concepts associated with the field of nanotechnology.
- You will be able to differentiate between the nanoscale, nanoscience, and nanotechnology.
- You will gain a greater understanding of the nanoscale, including the difference between linear and logarithmic scales, as well as the relative and absolute size of things.

II. Background Information

1. Introduction to the Field of Nanotechnology

Nanotechnology is quickly becoming a more familiar term to many people. Although many may have heard the term used in movies such as *Terminator 3: Rise of the Machines* or in books such as *Prey*, very few people could tell you what nanotechnology is or why it is important to understand. This unit will introduce the vast field of nano, as well as explore the key terms, definitions, and distinctions of this rapidly growing area of study.

Some people in the field of nanotechnology have suggested that it might be the next industrial revolution. Nanotechnology deals with the small, but its impact on areas such as manufacturing, medicine, and defense could be quite large. Nano is truly a multidisciplinary field, so to gain a better understanding of what it is and why it is important, the terminology associated with nano will be discussed.

Nanoscience is an umbrella term that covers many areas of research dealing with objects at the **nanoscale** (objects that are measured in nanometers). A **nanometer** (nm) is a billionth of a meter, or a millionth of a millimeter. With this in mind, **nanotechnology** can be defined as the purposeful manipulation of matter at the nanoscale or atomic level to achieve a defined goal. While many definitions for nanotechnology exist, the National Nanotechnology Initiative (NNI), the authority on nanoscale technology, research, and development, calls it *nanotechnology* only if it involves all of the following:

1. Research and technology development at the atomic, molecular, or macromolecular levels, in the length scale range of approximately 1 to 100 nanometer/s

2. Creating and using structures, devices, and systems that have novel properties and functions because of their small and/or intermediate size

3. The ability to control or manipulate on the atomic scale

For more information on this federal program established to coordinate the multiagency efforts in nanoscale science, engineering, and technology, see the Resource Index under *National Nanotechnology Initiative.*

2. The Nanoscale

Size is the one aspect that unifies all of the different areas of nano, the term itself is the metric prefix that means one-billionth; therefore, a nanometer is one-billionth of a meter. Figure 4-1 shows some examples of things that are measured in nanometers.

But just how small is one-billionth of a meter? You can easily estimate the length of a meter (m) by looking at the length of an adult's arm. It is roughly 3 feet, or 1 meter. In fact, most people can walk about 1,000 meters in about 15 minutes. But, when lengths get longer than what we can personally experience, such as the distance from the earth to the International Space Station, it often becomes difficult for people to understand. The same is true for small-sized objects. Our eyes give us the ability to see things as small in size as an ant's eye and the point of a pin (approximately 50 micrometers) and, as such, objects that are smaller than this become difficult for us to conceptualize.

To help us better understand this idea, we can look at some common size units and determine where the nanoscale is in relation to them (Table 4-1).

What if we measured the size of common objects in terms of nanometers? If we did, we would find that a red blood cell is about 10,000 nanometers across and the width of a human hair is 100,000 nanometers. The head of a pin is a million (1,000,000) nanometers wide and an adult man who is 2 meters tall (6 feet 5 inches) is about 2 billion (2,000,000,000) nanometers tall!

With these comparisons, you can begin to appreciate how small the scale is that nanoscientists and nanotechnologists use. Can you imagine that at the nanoscale, scientists are manipulating objects that are more than one-millionth the size of the period at the end of a sentence? For more information regarding orders of magnitude and the powers of ten, see the Resource Index under Orders of Magnitude and Interactive Tutorials.

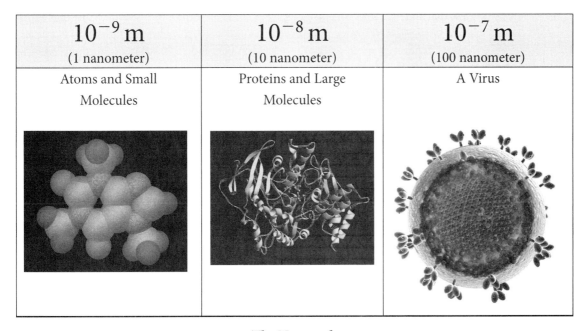

$10^{-9}\,\text{m}$	$10^{-8}\,\text{m}$	$10^{-7}\,\text{m}$
(1 nanometer)	(10 nanometer)	(100 nanometer)
Atoms and Small Molecules	Proteins and Large Molecules	A Virus

The Nanoscale

Figure 4-1 Representative Examples of Nanoscale Objects
(from left to right: SMC Images/Getty Images; Manuel C.; Dr. Peitsch/CORBIS; 3D4Medical.com/Getty Images)

Unit	English Expression	Magnitude as a Number	Magnitude as an Exponent	Relative Size
Meter	One	1	10^0	Length of an adult's arm
Centimeter	One Hundredth	0.01	10^{-2}	Width of a fingernail
Millimeter	One Thousandth	0.001	10^{-3}	Thickness of a dime
Micrometer	One Millionth	0.000001	10^{-6}	Diameter of a single cell
Nanometer	One Billionth	0.000000001	10^{-9}	Length of 10 hydrogen atoms lined up

3. Scales and the Size of Things

Scale is the relationship between an actual measurement and the way that measurement is represented numerically or visually. A scale usually has a succession of ascending and descending steps, or relative dimensions, used to assess the **absolute** or **relative size** of some property (e.g. length, temperature, or mass) of an object. We use many different types of scales, but two of the more common are **linear** and **logarithmic**. In a linear scale, the lengths represented between each of the marks is equal. So, in Figure 4-2, the distance between 1 and 2 is the same as the distance between 3 and 4.

Scales can range from smaller than an atom, as we have already seen, to larger than the universe. Therefore, a linear scale may not be an accurate or convenient way to represent such enormous size differences. Often a logarithmic scale uses the orders of magnitude, or the powers of 10, to represent and compare the relative sizes of objects with actual lengths. The pH scale in Figure 4-3 is an example of a logarithmic scale.

The pH of pure or distilled water is 7. If you move up (in the direction of increasing acidity), you see that lemon juice has a pH of about 2 and sulfuric acid has a pH of about 1. If the pH scale were a linear scale, such as the ruler in Figure 4-2, one might think that sulfuric acid was just slightly more acidic than lemon juice. But that is not the case. Since the pH value is on a logarithmic scale, we see that sulfuric acid is actually about 17,000 times more acidic than lemon juice. Figure 4-4 depicts another example of a logarithmic scale that represents our world, from enormous galaxies to tiny atoms.

Figure 4-2 Ruler as an Example of a Linear Scale

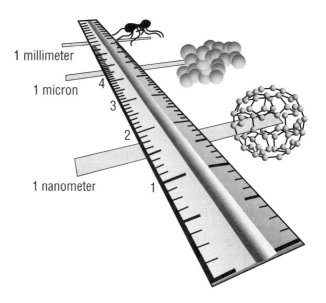

1 millimeter

1 micron

1 nanometer

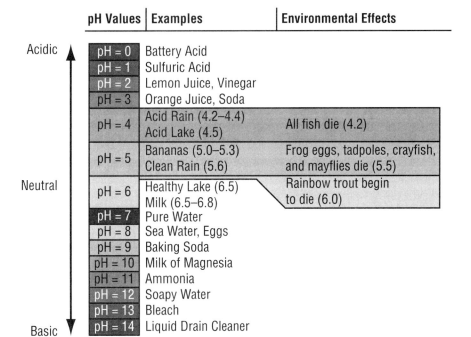

Figure 4-3 The pH Scale as an Example of a Logarithmic Scale

pH Values	Examples	Environmental Effects
pH = 0	Battery Acid	
pH = 1	Sulfuric Acid	
pH = 2	Lemon Juice, Vinegar	
pH = 3	Orange Juice, Soda	
pH = 4	Acid Rain (4.2–4.4) Acid Lake (4.5)	All fish die (4.2)
pH = 5	Bananas (5.0–5.3) Clean Rain (5.6)	Frog eggs, tadpoles, crayfish, and mayflies die (5.5)
pH = 6	Healthy Lake (6.5) Milk (6.5–6.8)	Rainbow trout begin to die (6.0)
pH = 7	Pure Water	
pH = 8	Sea Water, Eggs	
pH = 9	Baking Soda	
pH = 10	Milk of Magnesia	
pH = 11	Ammonia	
pH = 12	Soapy Water	
pH = 13	Bleach	
pH = 14	Liquid Drain Cleaner	

Acidic / Neutral / Basic

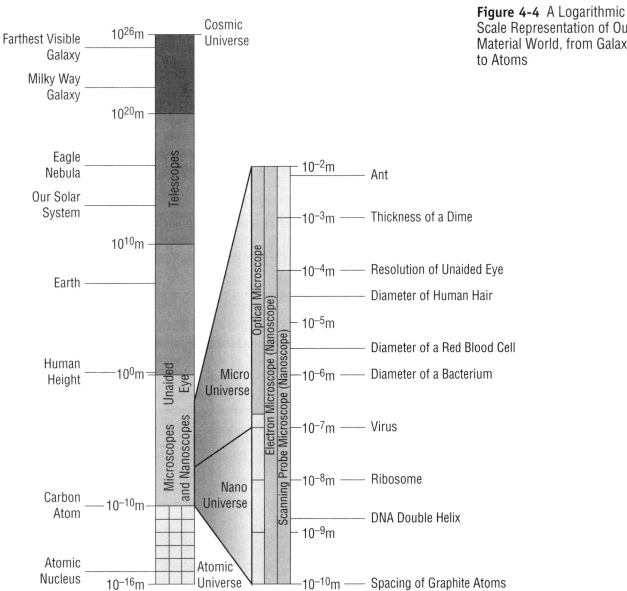

Figure 4-4 A Logarithmic Scale Representation of Our Material World, from Galaxies to Atoms

nanoscience	nanotechnology	relative size
nanometer	scale	linear scale
nanoscale	absolute size	logarithmic scale

III. Student Materials

Use Project 1 Design Brief: A Sense for the Nanoscale to complete the project.

UNIT 4: PROJECT 1   DESIGN BRIEF: A SENSE FOR THE NANOSCALE

In this project, you will conduct a two-part experiment in which you will use multiple sensory systems (taste and vision) to conceptualize the nanoscale.

Nano means one billionth. Nano is too small to see with the human eye or with optical microscopes. To help gain a perspective of the nanoscale, you are going to use more than just your visual senses to understand how small nano really is.

Introduction

- Using a meter stick, measure 1 meter in length on the floor. Place tape at each end so that you have exactly 1 meter in length between the two pieces of tape.

- There are 1 billion nanometers in a meter. Looking at 1 meter, it is hard to image lining up 1 billion objects in between those pieces of tape. Think of how thin paper is. If you tried to stack paper up equal to 1 meter in height, you wouldn't even come close to having 1 billion pieces of paper.

- Trying to imagine how small nano is can be tough, especially since we can't see it with the naked eye.

Since our visual system isn't sensitive enough to do the job, maybe our taste or chemical senses are more sensitive. Can we taste nano? We are going to conduct a taste test to determine if we can taste nano. We will do two versions of the experiment to try to gain some perspective of the nanoscale.

Experiment 1

Overview

For this experiment, you will take a substance (lemon juice) and go through a series of reducing concentration levels until you reach a solution that has 1 part in 1 billion lemon juice.

Materials Needed to Conduct This Experiment

- 10 clear plastic cups
- 1 bottle of lemon juice
- Pitcher of drinkable water
- Teaspoons
- Paper and pencil to take notes about experimental findings
- Graphing paper or graphing software

Experiment 1 Procedure

1. Arrange 10 plastic cups in a row and label them numerically.
2. Place 10 teaspoons of lemon juice in the first cup.
   - This cup represents a 100 percent concentration of the substance.

3. Remove 1 teaspoon of lemon juice from cup #1 and place it into cup #2 and add 9 teaspoons of water to cup #2.

  • The cup has 10 teaspoons of solution, 9 parts water, 1 part lemon juice. This cup represents 1 part lemon juice out of 10 parts (1/10).

4. Remove one teaspoon of lemon juice from cup #2 and add it to cup #3. Then add 9 teaspoons of water.

  • This cup is 1/100 lemon juice.

5. Continue to remove 1 teaspoon and add it to the next cup in addition to adding 9 teaspoons of water.

6. When you reach cup #10, you will have reduced the concentration from 100 percent concentration to one billionth (1/1,000,000,000) concentration.

7. Use a 5-point scale and rate the intensity of the lemon taste for each cup.

  • 1 = no lemon taste

  • 2 = weak lemon taste

  • 3 = mild lemon taste

  • 4 = moderate lemon taste

  • 5 = strong lemon taste

While Conducting the Experiment

  • Take notes of any problems encountered that might affect the results.

  • Record the rating of lemon intensity for all 10 cups.

  • Graph the data to convey at what concentration level lemon juice is no longer detected using taste.

Experiment 2

Overview

Now that you have seen whether you can taste nano, you will look at nano by using two senses together. Experiment 2 is similar to Experiment 1, but you will use dark fruit juice (e.g., fruit punch or grape juice). The fruit juice allows you to test your taste or chemical system, while also allowing you to test your visual system.

  Materials Needed to Conduct This Experiment

  • 10 clear plastic cups

  • 1 bottle of fruit punch or grape juice

  • Pitcher of drinkable water

  • Teaspoons

  • Paper and pencil to take notes about experimental findings

  • Graphing paper or graphing software

Experiment 2 Procedure

1. Arrange 10 plastic cups in a row and label them numerically.

2. Place 10 teaspoons of fruit juice in the first cup.

  • This cup represents a 100 percent concentration of the substance.

3. Remove 1 teaspoon of fruit juice from cup #1 and place it into cup #2 and add 9 teaspoons of water cup #2.

  • The cup has 10 teaspoons of solution: 9 parts water, 1 part fruit juice. This cup represents 1 part fruit juice out of 10 parts (1/10).

4. Remove 1 teaspoon of fruit juice from cup #2 and add it to cup #3. Then add 9 teaspoons of water.
    - This cup is 1 /100 fruit juice.

5. Continue to remove 1 teaspoon and add it to the next cup in addition to adding 9 teaspoons of water.
    - When you reach cup #10, you will have reduced the concentration from 100 percent concentration to one billionth concentration.

6. Use a 5-point scale and rate the intensity of the fruit taste for each cup.
    - 1 = no fruit taste
    - 2 = weak fruit taste
    - 3 = mild fruit taste
    - 4 = moderate fruit taste
    - 5 = strong fruit taste

7. Use a 5-point scale and rate the intensity of color for each cup.
    - 1 = no color
    - 2 = weak color
    - 3 = mild color
    - 4 = moderate color
    - 5 = strong color

While Conducting the Experiment

- Take notes of any problems encountered that might affect the results.
- Record the rating of *intensity* and *color* for all 10 cups.
- Graph the data to convey at what concentration level fruit juice is no longer detected using taste or vision.

Follow-up

More than likely you will find that you can easily taste and see the substance in the first cup that is 100 percent concentrated. But after the third or fourth cup, you will likely no longer be able to taste or see any indication of the substance.

After the third or fourth cup when you can no longer see or taste anything, you begin to develop a sense for just how far removed the nanoscale really is from what we know and are used to working with.

Look at the tenth cup. If you couldn't see or taste beyond the third cup, imagine how difficult it is to actually be able to see and work at the level of that tenth cup, which represents the nanoscale.

What types of things could you do so that you could get useful information from that tenth cup? Could you build an artificial or bionic eye capable of seeing information that is normally invisible to the human eye? Could you build better taste receptors on the tongue so that the taste system would be capable of detecting lemon juice in a 1/1,000,000,000 concentration level? These are some modifications that might allow us to reach that tenth cup. In reality, tunneling microscopes and nanomanipulators represent the technology that enables researchers to actually work in that tenth cup.

Design Constraints (Things You Must Do)

You will create a lab report in the form of a booklet that details the experiments and results from the A Sense for the Nanoscale Design Brief. The completed lab report should include:

- An Introduction section explaining the purpose of the laboratory experiments.

- A detailed description of what was done in the Methods section.
- A Results section that includes the observations that were made and the ratings of the intensity and/or color for all 10 cups (for each part of the experiment).
- The Results section should contain a table(s) that indicates the concentration levels at which sight or taste was still useful.
- The Results section should also contain graphs of the data to convey at what concentration level liquids were no longer detected using taste and/or vision.
- A Discussion section that explains the results and any problems encountered that might have affected the results.
- For concentration levels that were undetectable using sight or taste, use the Internet and find technology that is used at that level.
  - For example, micro means 1/1,000,000 or 10^{-6}. A type of technology that uses the microscale is a capacitor. A capacitor is a device used to store electrical charge in a circuit. Electrical charge or capacitance is measured in farads. One farad represents a very large amount of charge. One farad equals 6,280,000,000,000,000,000 electrons. This amount is hard to work with and so capacitors are rated using microfarads that change the overly large unit into a unit that is easier to work with, 1 μF = 0.000,001 F.
- A Conclusion section that summarizes the experiments and links back to the nanoscale.

PROJECT 2
What Is Nanoscience?

I. Project Lesson Plan

1. Project Description

During this project, you will explore the multidisciplinary field of nanoscience more deeply and learn how nanoscientists plan to build with atoms. In doing so, you will be introduced to the forces and effects that take place at the nanoscale. Additionally, you will discover the way in which the characteristics and properties of matter differ when one moves from the macroscale to the nanoscale.

2. Learning Objectives

- You will gain a basic understanding of atoms and their unique properties as they serve as the building materials for nanoscientists.
- You will describe some of the properties and characteristics of matter at the nanoscale.
- You will compare and contrast forces and effects that dominate at the macroscale and nanoscale.

II. Background Information

1. Building with Atoms

Nanoscience is dependent on the ability to measure, design, use, and construct at the **nanoscale** level. At a much larger scale, woods and metals are used as building materials to construct objects such as houses or computers. At the nanoscale, the building materials used are on the **atomic scale**.

Atoms Versus Molecules

An **atom** is the smallest unit of a chemical element. Atoms are the substance of matter. Chairs, desks, books, and everything around us is made up of atoms. Atoms can combine to form new structures. A **molecule** consists of two or more atoms held together by a chemical bond. For example, two oxygen atoms can combine to form a molecule of O_2. See Figure 4-5 for the difference between an atom and a molecule. A **compound** is a substance made up of atoms from at least two elements. For example, two hydrogen atoms can combine with an oxygen atom to form the molecular compound H_2O.

Building with Atoms and Molecules

Atoms have three basic components: **electrons, protons**, and **neutrons**. Electrons have a negative charge, protons have a positive charge, and neutrons have a neutral charge. Atoms are combined with one another to form molecules by developing chemical bonds through sharing electrons. The molecule is the smallest unit, or configuration, of atoms that can exist by itself. For a water molecule, this is two hydrogen atoms and one oxygen atom. Sharing electrons between two atoms creates a **covalent bond**. See Figure 4-6 for an illustration of covalent bonds in water.

Closely related to covalent bonds are **ionic bonds**. These types of bonds happen between atoms that have lost or gained one or more electrons and therefore

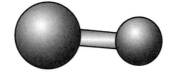

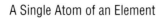

A Single Atom of an Element   A Molecule Made Up of Elements

Figure 4-5 Atom Versus Molecule.

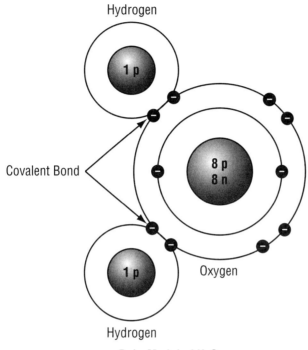

Hydrogen

1 p

Covalent Bond

8 p
8 n

1 p

Oxygen

Hydrogen

Bohr Model of H$_2$O

Figure 4-6 Covalent Bond Two hydrogen atoms combine with an oxygen atom to form a water molecule by sharing electrons.

have developed an unbalanced charge. These atoms are now called **ions**. Since electrons are negatively charged, extra electrons will make the atom negatively charged, while a shortage of them will make the atom positively charged. One way for the atom to stabilize is to regain or lose the additional electrons and balance its charge. Alternatively, it can join with another atom that has the opposite charge imbalance, creating an ionic bond. Ionic bonds involve transferring electrons between two or more atoms. For example, sodium that has lost an electron can join with chloride that has gained an extra electron to create common table salt, sodium chloride. Ionic bonds can also go beyond creating a single molecule to making larger crystalline structures. A grain of table salt is, at the nanoscale, a massive crystal structure of sodium and chloride atoms.

Molecules that are constructed with covalent bonds can also have electromagnetic properties similar to ions. When a molecule is made up of different types of atoms, these atoms exert differing amounts of "pull" on the electrons, keeping more or fewer electrons near each type of atom. If these atoms are arranged in the molecule in a *nonsymmetric* fashion, then electrons will be nonsymmetrically arranged around the molecule.

In this case, a line, or axis, can be put through the molecule that has an imbalance of charge along it. That is, one end of the molecule will be more negatively charged than the other end, creating a **dipole moment**. This is the case for water molecules (Figure 4-6), where the oxygen end of the molecule is more negatively charged than the hydrogen end. Because the water molecule is nonsymmetric, it has a dipole moment. As water molecules are grouped together, the more negative oxygen end aligns with the hydrogen end of a neighboring molecule, creating a **hydrogen bond**. Though hydrogen bonds are not as strong as the covalent bonds that hold the water molecule together, they are strong enough to give groups of water molecules interesting properties. This will be explored in more detail in the design brief activity.

Metal atoms, such as gold, copper, and platinum, give up electrons easily, becoming positively charged ions. In **metallic bonds**, the electrons are held even more loosely than in ionic bonds. In the crystalline lattices of metals, electrons essentially float free between the positively charged metal ions. The nature of these bonds is the source of many of the properties observed in matter. These will be discussed in the next section. Nanoscience deals with building and manufacturing at this atomic level. To do so, nanoscience has to take advantage of the properties of atoms and molecules and the bonds that hold them together.

2. Properties of Matter

Characteristics and *properties* are essentially ways of describing different qualities and behaviors of a substance. Scientists have accumulated a vast amount of knowledge regarding the properties of various substances. Take, for example, a piece of metal such as gold. At this **macroscale**, we know its optical properties (such as color and transparency), electrical properties (such as conductivity), physical properties (such as density and hardness), thermal properties (such as boiling point and melting point), and chemical properties (such as its reactivity and reaction rates). As a result, we can use this information to predict what gold will do under different conditions and to make decisions about whether it is good material to use when we are building or synthesizing something. We know that because of metallic bonds, gold can let go of its electrons, making it a good conductor. Because the metallic bonds are weak, it melts at a low temperature and can be reshaped with blows of a hammer. Crystalline salts, held together with ionic bonds, usually melt at much higher temperatures and would shatter if hit with a hammer.

However, at the nanoscale level, objects do not necessarily act the way one would expect them to at the macroscale. Properties such as conductivity, hardness, and color act differently. Nanoscientists aim to provide a basic understanding of how the characteristics and properties of matter change as their size gets smaller and smaller.

Considering again our example of gold, if one were to take a piece of gold a few centimeters in size, cut it in half, then cut one of the halves in half and continue cutting each half until the piece of gold had been reduced from centimeters down to millimeters, the gold would still retain the same properties as it did when it was in one large piece.

However, if the gold continued to be cut into smaller and smaller pieces to reach the size of nanometers, the color of gold would change. What we see as color results from the quantity and quality of lightwaves entering our eyes. How far the light wave penetrates into the surface of matter (such as gold) and how it interacts with molecules determines the characteristics of the light wave when it bounces back out and enters our eyes. How gold atoms arrange themselves when there are only a few of them and how light waves interact with them is much different than when gold atoms are in a macroscale chunk. Nanoscale gold particles can be purple, orange, red, or even green depending upon their size and shape.

Changes in gold's color at the nanoscale level are due to the fact that the material that was once bulk has been cut down to only a few atoms. These atoms behave differently when they aren't grouped into a large quantity. Once the atoms are again grouped into a bulk quantity, such as at the macroscale level, they will begin to combine. At that point, the color that is commonly associated with gold will return.

3. Changes at the Nanoscale

In addition to such changes in optical properties (e.g., color, luster, fluorescence), the electrical (e.g., conductivity) and mechanical (e.g., strength and flexibility) properties of nanosized matter may change. But what is the reason for such changes? There are four main factors or forces that are different when we consider nanosized particles of substances. First, due to the extremely small mass of these particles, *gravitational forces* are almost non-existent. Instead *electromagnetic forces* dominate and ultimately determine the behavior of atoms (and molecules). Second, at nanoscale sizes, we need to use *quantum mechanics* to describe the particle motion and energy transfer instead of the classical mechanics descriptions. Third, nanosized particles have a very large *surface area-to-volume ratio*. And fourth, at this size, the influences of random thermal molecular motion play a much greater role than they do at the macroscale.

The design brief for this unit provides an example of how electromagnetic forces begin to dominate over gravitational forces as you get closer and closer to the nanoscale. Think about hanging a weight on a thin string. The string stays together due to the strength of the individual fibers and their ability to "stick" to each other. As more weight is hung onto the string, the gravitational force pulling on this mass finally overcomes the strength of the string, and the string breaks. Similarly, at the macroscale, the gravitational forces often overcome electromagnetic forces between molecules.

Instead of thinking of gravitational force acting on a linear string, you can think of it acting on a surface or "skin." Imagine a thin paper bag as large as your house and start filling it with sand. You can imagine that at some point, the weight of the sand in the bag will become too great and the bag will rupture. What if you made the bag half as large? Would it rupture before it is filled all of the way? Probably, but at some point if you continued on with the experiment, the bag would become small enough that it could support the weight of the sand inside of it. Water acts the same way, except that the "skin" on the water is the same as what is inside. A lattice of water molecules covers, the outer surface of the water and is held together by electromagnetic forces. In large quantities, the weight of the water overcomes the electromagnetic forces and the water ruptures and flows— just as it does when you pour a pitcher of water on the floor. However, when the quantity of water gets small enough, the skin of the water drop is stronger than the gravitational forces pushing outward, holding the drop together.

Water is notable because of the presence of hydrogen bonds between water molecules. Although these hydrogen bonds are relatively weak compared to the chemical bonds within a molecule, they are of sufficient strength to make water unusually *cohesive*. It is water's cohesiveness that allows it to form a nearly spherical "bead" when a single drop is placed on a flat, nonporous surface (Figure 4-7). Water is also a polar molecule, that is, it has a negatively charged end and a positively charged end. The negative end of one water molecule is attracted to the positive end of another, and they "stick together" tightly. When you drop water on a coin, the molecules of water in the new drop stick to the water droplets in the old drops and form a dome shape as shown in Figure 4-7.

This cohesiveness gives water its high degree *of surface tension*, which is visible in the small indentations made by the legs of certain insects that can literally walk on water. The perimeter of a water strider's foot that is in contact with the water surface is a function of the foot's surface area ($L2$). The surface force of the water tension depends on the insect's foot perimeter. But the weight to be supported by the surface force depends on the volume of the insect. Weight is directly dependent upon gravity, which in this case is not much, since the insect is so small.

Rubbing alcohol, on the other hand, is a mixture consisting of 70 percent isopropyl alcohol and 30 percent water. It does contain some hydrogen bonds within

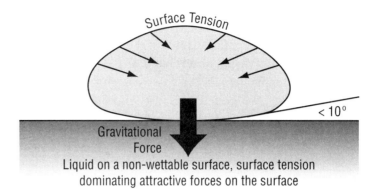

Figure 4-7 Surface Tension Surface tension dominates over gravitational force to keep the water in a bead.

Surface Tension

Gravitational Force

< 10°

Liquid on a non-wettable surface, surface tension dominating attractive forces on the surface

its structure, but not nearly as many as occur in pure water. Rubbing alcohol will form a bead when dropped onto a flat, nonporous surface, but the bead will be slightly flatter and larger in diameter than a corresponding bead of pure water. The design brief will explore this phenomena.

Key Terms

nanoscience	electron	ion
nanoscale	proton	dipole moment
atomic scale	neutron	hydrogen bond
atom	covalent bond	metallic bonds
molecule	ionic bond	macroscale
compound		

III. Student Materials

Use Project 2 Design Brief: Observing the Effects of Size on the Properties of Matter to complete the project.

UNIT 4: PROJECT 2 DESIGN BRIEF: OBSERVING THE EFFECTS OF SIZE ON THE PROPERTIES OF MATTER

Design Constraints: Things You Must Do

Below are constraints, or guidelines, to follow during this design brief:

1. Work in pairs to conduct these experiments.

2. Follow all procedures and instructions carefully.

3. Record all observations and results as you conduct the experiments.

4. You will create a lab report that details the experiments and results from each part of the Design Brief. The completed lab report must include:

   • An Introduction section explaining the purpose of the laboratory experiment.

   • A Methods section with a detailed description of what was done.

   • A Results section that includes sketches and notes of observations that were made. The Results section should contain a table(s) and graph(s) of the data.

   • A Discussion section that explains the results and any problems encountered that might have affected the results.

   • A Conclusion section that summarizes the experiments and links back to the nanoscale.

Student Materials

Small plastic cups
Water
Rubbing alcohol
Disposable pipettes
Pennies
Paper towels

Safety Note: Be sure you do not taste the alcohol or get the alcohol in your eyes.

Student Procedures

Gather the materials your group will need. Your group will need a plastic cup containing about a half-inch of water and a plastic cup with an equal amount of rubbing alcohol. Also get one penny, two paper towels, and one disposable pipette for each person in the group.

Each member of the group should complete the following steps:

1. Take one of the pennies and place it head side up on a paper towel on the lab table in front of you.

2. Slowly and gently, use the pipette to place drops of water on the penny until one drop makes the water overflow the edges of the penny. Be sure to count the drops as you add them. **Record the number of drops you added minus the one that made the water overflow**. This data will be needed for your final report.

3. Dry the penny thoroughly with the other paper towel and squeeze all the water out of the pipette. Repeat steps 1 and 2 two more times. Each student should conduct and record the data from three trials.

4. Repeat the same procedures (steps 1 through 3) with the rubbing alcohol.

Student Questions

1. Did you get different averages for the two liquids? If so, which liquid allowed you to put more drops on the penny?

2. In your group, discuss why you think the number of drops were different. Try to form a hypothesis to explain the differences on which everyone in the group agrees.

PROJECT 3
Tools of Nanotechnology

I. Project Lesson Plan

1. Project Description

You will learn about the tools that enable nanotechnologists to research, design, and nanoengineer at the nanoscale. This project begins with a description of the relationship between nanoscience and nanotechnology. Next, you are introduced to the common tools and techniques that scientists in the field of nanotechnology use in their work. You will use current technology to generate an image by using a procedure similar to that of scanning probe microscopy.

2. Learning Objectives

- You should be able to understand the relationship between nanoscience and nanotechnology.
- You will become familiar with the components of the scanning tunneling microscope and the atomic force microscope.
- You will develop an understanding of the process that scanning probe microscopes use to produce images of atomic scale objects.
- You will be able to discuss the differences in the underlying technology of an atomic force microscope (AFM) and a scanning tunneling microscope (STM).

II. Background Information

1. Introduction to the Field of Nanotechnology (Review)

Nanotechnology is becoming an increasingly familiar term to many people. Although many may have heard the term used in movies such as *Terminator 3: Rise of the Machines* or in books such as *Prey*, very few people could tell you what nanotechnology is or why it is important to understand it. This unit will introduce the field of nanotechnology, as well as some of the key concepts and tools, as well as the potential and current uses of nanotechnology.

Nanotechnology refers to creating some material or machine through the manipulation of individual atoms or molecules. Some people working in this emerging field have suggested that it might well be the next Industrial Revolution. Nanotechnology deals with the small, but its impacts on areas such as manufacturing, medicine, and defense could be quite large.

2. The Connection between Nanoscience and Nanotechnology

Nanoscientistis study the fundamental physical, chemical, and biological properties of objects (e.g., atoms, molecules, and structures) at the nanoscale. Nanotechnologists apply this knowledge to advance technology. Therefore, nanotechnology is the application of nanoscience at the atomic or molecular scale of approximately 1 to 100 nanometers. It consists of the tools and techniques used to produce nanoscale objects and devices.

For example, a nanoscientist studies the fundamental or underlying properties of **carbon nanotubes**. A nanotechnologist designs and develops uses for carbon nanotubes, such as using them to build stronger and faster military equipment or medical devices.

It is difficult to imagine having the ability to manufacture structures atom by atom, but that is exactly what researchers in nanotechnology hope to do. Researchers in the field of nanotechnology come from a variety of disciplines such as materials science, electronic engineering, mechanical engineering, and medicine. There are two approaches regarding how to design and develop in nanotechnology: *top-down* and *bottom-up*. Researchers using the **top-down approach** are developing machining and etching techniques to create nanoscale structures. This is analogous to taking a larger bulk substance and whittling away the substance until you reach the nanoscale. In contrast to this approach, researchers who adhere to the **bottom-up approach** are investigating ways to build and manufacture structures, such as molecular machines, molecule by molecule or even at the level of atom by atom. It may seem like it would be impossible to build using a bottom-up approach, but there are examples of naturally occurring molecular machines that do indeed build at the molecular level.

3. Issues in Nanotechnology Manufacturing

A few of the proposed uses and applications for nanotechnology include **nanorobots** used as medical delivery systems and military fatigues equipped with **nanosensors** that can detect biotoxins. However, the technology is not yet advanced enough to begin manufacturing and mass producing at the nanoscale level.

Nanotechnology deals with measuring, designing, and constructing at the nanoscale level. However, mass production technology currently relies on macroscale level tools. The first issue with nanotechnology production is being able to grasp and manipulate nanoscale size objects. Therefore, the goal of nanotechnologists is to create **molecular assemblers**, which are machines that are molecular in size. These machines would be used to assemble additional assemblers as well as to build other small structures or objects.

The structure of an object would be based on the way the atoms were arranged, similar to how the shape of a house is based on the way the walls are put together. On a large scale, nails, glue, staples, and cement are used to hold together the various building materials. At the nanoscale, the atoms are held together by chemical bonds. Therefore, the shape of nanoscale objects is determined by the arrangement of atoms and whether they can be joined by a chemical bond. See the Resource Index section under Molecular Assemblers for an interesting debate between two of nanotechnology's leaders, K. Eric Drexler and Richard E. Smalley, on the potential of molecular assemblers.

Another challenge to working with the nanoscale is the fact that the nanoscale deals with a dimension that ranges from hundreds to only tens of nanometers. The nanoscale is so small that objects at that scale are not even visible through a microscope. Work at the nanoscale requires tools that can enable individuals to see, manipulate, and engineer at the nanoscale level. New techniques in microscopy are providing researchers with tools that enable them to do just these types of activities.

4. The Tools of Nanotechnology

To make sense of the world around us, we rely on our sensory system. Sight, sound, smell, and touch help us make sense of the stimuli that surround us. When stimuli are smaller than our unaided senses are capable of detecting, technology helps to extend our sensory capabilities. Similar to how hearing aids help people hear frequencies no longer detectable by their auditory nerves, optical microscopes help people see microscopic objects.

Nanostructures are smaller than microscopic objects, so they are too small to be seen with traditional optical microscopes. The wavelength of visible light ranges between 400 and 700 nanometers. Since nanostructures can be as small as only a few nanometers, using microscopy techniques that depend on light waves for imaging won't work. The nanoscale represents the range where objects are 10^{-9} meters in size, which is smaller than the resolution capabilities of optical microscopes.

So how do we see nanostructures if we can't really "see" them? **Scanning probe microscopy (SPM)** makes seeing objects in the nanoscale possible (Figure 4-8). Scanning microscopes map the surface of an object, and a computer then creates a visual image based on data as measured or detected by the microscope. There are several types of scanning probe microscopes and each type creates an image by measuring slightly different (i.e., not necessarily touch) properties. These techniques are so sensitive that they can produce images of individual atoms.

The **scanning tunneling microscope (STM)** was invented by Gerd Binning and Heinrich Rohrer at IBM in 1981. This invention later won them the Nobel Prize in Physics. The STM can generate images with atomic resolution. The technique to produce these images relies on the quantum mechanical effect known as **tunneling**. Images are created in scanning probe microscopy by a computer rather than by using the human eye. The computer creates an image by mapping out data points. The data points the computer maps use are from interaction between the microscope and the object being scanned.

In STM, the data points are based upon tunneling, which involves an exchange of electrons between atoms. Since atoms are the basis of all matter, the tip of the microscope is a series of atoms, and the sample being scanned is also composed of atoms. These atoms will interact with one another through tunneling and it is this interaction that provides the data points from which the computer generates an image.

Let's examine the components of the STM to gain a better understanding of how the microscope is used to produce an image of something so small. The main component of a scanning tunneling microscope is a probe that is similar to a sharp tip (Figure 4-9). To use the STM to detect individual atoms, the tip itself is very small. In fact, the tip tapers down to a point that is only a single atom across. It is the interaction between the atoms that constitute the tip and the atoms that constitute the sample that provides the data points the computer uses to create a topographical map of the surface of the sample.

The tip interacts with the surface of the sample to create the map of the surface without changing its appearance. The tip is a conductor of electricity and as long as the sample is also a conductor of electricity, then the two can interact through tunneling. When the tip is close enough to the sample, about 1 nanometer away, electrons from the atoms on the tip will begin to travel to the atoms that constitute the sample and vice versa. A fixed bias voltage is applied to the tip and a feedback loop is used to control

Figure 4-8 Components of a Scanning Probe Microscope Scanning probe microscopes provide data points based on mapping the surface of an object and having a computer create a visual image based on the data points.

(U. Bellhaeuser/Getty Images)

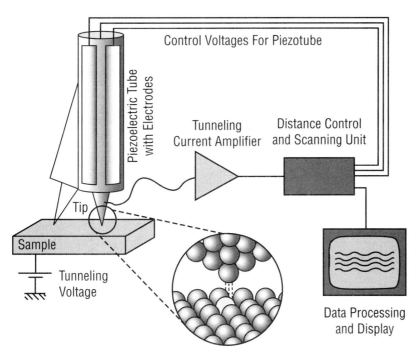

Figure 4-9 How a Scanning Tunneling Microscope Works

(Courtesy of Michael Schmid, IAP/TU Wein, Austria)

the voltage so that a constant tunneling current can be maintained. When the tip scans over a sample, sometimes the surface contains electron clouds that readily exchange electrons with the tip. When this facilitation in conduction occurs, the tip moves upward in an effort to maintain the desired constant current. When the tip scans over an area with an electron cloud that does not favor conduction, the tip moves downward to reduce the resistance and achieve the desired constant current.

The computer detects these changes in tip movement and maps them to provide a topographical map of the surface of the sample. The amount of adjustment provides a number, and the computer can then graph the data by plotting it onto a grid. The amount of adjustment represents the height of that point on the grid. The computer generates the topographical map in grayscale but other features such as curvature or differences in height across the grid can be used as a basis for adding color to the image. Figure 4-10 is an example of using a STM to arrange iron atoms at the nanoscale.

The movement of the tip up and down and the scanning movement across the sample are very small movements so that they can detect individual atoms. Let's

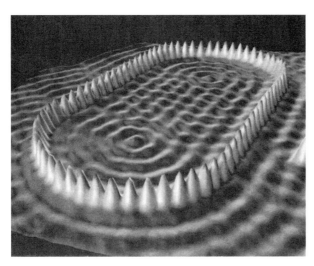

(Image reproduced by permission of IBM Research, Almaden Research Center)

Figure 4-10 Sample Image from a High-Vacuum, Low-Temperature Scanning Tunneling Microscope (STM) The pictured quantum corral is a stadium-shaped ring of 76 iron atoms, each moved into position on a copper surface by a scanning tunneling microscope. It is 285 by 141 angstroms in size.

look at the part of the STM that is responsible for controlling the nanoscale movements of the tip.

Tunneling microscopes, such as the STM, work by measuring changes in the electrical current between the probe and the sample. Force microscopes, such as the **atomic force microscope (AFM)**, work by measuring a magnetic or mechanical force on the surface of the sample (Figure 4-11). The AFM's tip actually touches the sample. If the tip works by maintaining a constant force while scanning over the surface, it is called **contact mode**. Another way the tip can work is by the tip having intermittent contact across the surface, which is called **tapping mode**.

AFMs work by scanning a sharp tip over a surface. The tip on the AFM is connected to a cantilever that deflects or moves as the tip drags over the surface of the sample. The AFM also uses a **piezoceramic tube scanner** with a feedback loop that detects tip deflection and then uses that signal to maintain constant tip force on the sample by adjusting the height of the sample as the tip scans over it. This is similar to the way the STM uses the piezoceramic tube scanner to control the tip height to maintain a constant tunneling current.

Another component of the AFM is the laser source, which has a beam that is reflected off the top of the cantilever to detect tip deflection as depicted in Figure 4-11. When the tip is deflected as it moves along the surface of the sample, the laser beam position changes. The laser beam reflects off the top of the cantilever onto a photodetector that consists of two photodiodes that are side by side. The difference between the two indicates the position of the laser beam and represents the angular deflection of the cantilever.

The design of the AFM provides two sources of data the computer can use to create an image. The feedback signal can be used to image the sample, or the tip deflection as detected from the laser beam can also be used. The AFM is held in place and it is the sample that is moved back and forth underneath the tip. Keeping the cantilever in place is what allows the laser source to reflect its beam off the cantilever. See the Resource Index under "Scanning Probe Microscopy Techniques" for more information on the tools used by nanotechnologists.

Figure 4-11 The Tip-Sample Interaction in Atomic Force Microscopy and Its Implications for Biological Application A tip is attached to a cantilever and tip deflections are detected using a laser source.

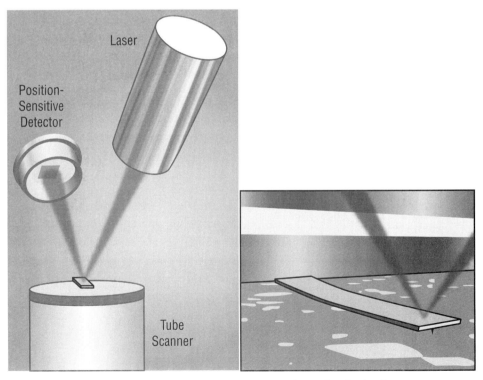

Laser

Position-
Sensitive
Detector

Tube
Scanner

(Copyright 1993 PhD thesis by Basett David R., California Institute of Technology)

nanotechnology
carbon nanotubes
top-down approach
bottom-up approach
nanorobots
nanosensors

molecular assemblers
scanning probe
   microscopy (SPM)
scanning tunneling
   microscope (STM)
tunneling

atomic force microscope
   (AFM)
contact mode
tapping mode
piezoceramic tube
   scanner

III. Student Materials

Use Project 3 Design Brief 1: Replicating Scanning Tunneling Microscopy and Project 3 Design Brief 2: Building an Atomic Force Microscope (ATM) Model to complete the projects.

UNIT 4: PROJECT 3   DESIGN BRIEF 1: REPLICATING SCANNING TUNNELING MICROSCOPY

In this project, you will collect data about the flow of current to produce an image of the surface of an object. The process you will use to collect data and produce an image is similar to how a scanning tunneling microscope (STM) works.

Design Constraints: Things You Must Do

Below are constraints, or guidelines, to follow during this design brief:

1. Work in pairs to conduct this experiment.

2. You will need:

    a. Piece of metal to act as a conductor of electrical current. You can use a piece of sheet metal (be careful of sharp edges) such as a metal face plate, or a metal serving tray.
    b. A pair of dark pantyhose or trouser socks
    c. Continuity tester
    d. Transparent tape
    e. Grid paper
    f. Pencil

3. Each person must decide which role they will assume first: experimenter or participant.

4. The experimenter will cut several pieces of tape and create a design on the surface of the metal plate. Make sure the participant does not see your design.

5. The experimenter will now cover the metal plate by stretching a pair of dark pantyhose or dark trouser socks over the metal plate.

6. The participant will take the continuity tester and some grid paper and will begin scanning the metal plate in a systematic fashion (from left to right in rows).

7. On the grid paper, map out the area on the metal plate in which you do not get a reading. The area where you get a reading is where metal is contacting metal. The area where you do not get a reading is where the probe is contacting the tape, which is the design the experimenter created. It is this data that you will use to produce an image of the design on the surface of the metal.

8. Once you have finished mapping out the surface based on the reading, compare your image to the actual design.

9. Switch roles, create a new design, and conduct the experiment a second time.

10. At the completion of the experiment, each person in the pair will have served in each role (experimenter and participant) and two different images will have been produced.

11. Note the ways in which this procedure is similar to and different from true scanning tunneling microscopy.

UNIT 4: PROJECT 3   DESIGN BRIEF 2: BUILDING AN ATOMIC FORCE MICROSCOPE (AFM) MODEL

In this project, you will construct a model of an atomic force microscope (AFM) by using a model building system. The goal will be to have your model demonstrate the functional characteristics of an AFM. In addition, you will visualize a 3D surface using your AFM model.

Design Constraints: Things You Must Do

Below are constraints, or guidelines, to follow during this design brief:

1. Work in pairs to conduct this experiment.

2. You will need:
   a. A model building system such as Lego™, Technics™, or Kinex™
   b. A light source such as a laser pointer or penlight (caution: Observe safety precautions if you are using a laser light source)
   c. Power supply for the light source
   d. A small mirror or highly reflective surface
   e. Tape or rubberbands for the light source and mirror
   f. White paper temporarily held on a stiff backing (e.g., clipboard or taped to cardboard)
   g. Irregular surface for the AFM to "read" (this can be a piece from the model building system)

3. Research AFM functionality. Start with the background material and sources listed in the Resource Index.

4. Draw sketches of the key components of an AFM.

5. Decide how the key components are going to be assembled and attached in your model. In particular:
   a. How is the light source going to be attached and aimed?
   b. How is the mirror going to be attached?
   c. Does your model system have a piece that will be appropriate for the probe tip, or are you going to have to create one?

Your goal will be to:

1. Assemble a functional AFM prototype. The light source needs to be reflected on the white paper so that you can mark key points of the light reflection on the paper.

2. "Read" the surface of an object at least 2 inches by 3 inches by moving the surface in multiple passes under the probe tip.

3. Each pass of the surface under the tip will move the light across your white paper. Mark the key points showing the surface "profile." Each

pass of the surface under the probe tip will be drawn on its own piece of paper.

4. Take all of the surface profiles and assemble a 3D representation of the surface you "scanned." This can be done in a number of ways.

   a. Create a physical model of the surface by mounting the profiles on a stiff backing, cutting them out, and mounting them spaced out upright.
   b. Create an isometric drawing of the profiles.
   c. Create a 3D computer model of the profiles.

PROJECT 4
Ethical and Societal Concerns Related to Nanotechnology

I. Project Lesson Plan

1. Project Description

In this project, you will examine and report on the potential risks and benefits of advances in the field of nanotechnology. Nanotechnology has been promoted as the next technological revolution and it has the potential to produce new materials that are stronger and weigh less. It also has the potential to change both medicine and defense. But do breakthroughs in nanotechnology present unique health and environmental dangers that need to be studied?

As with any new technology, there are several concerns regarding its impact in the areas of health, environment, privacy protection, the workforce, and the economy. This project culminates in a design brief in which you research the risks and benefits of this emerging technology. You design, conduct, and report upon a survey aimed at assessing the general public's level of understanding and opinions regarding nanotechnology.

2. Learning Objectives

- You will gain a basic understanding of the impact of nanotechnology on the areas of health, the environment, privacy protection, the workforce, and the economy.

- You will examine specific ethical and societal concerns associated with nanotechnology.

- You will conduct research to compare nanotechnology's potential good to its potential harm.

- You will design and conduct a survey intended to assess the general public's opinions on the risks and benefits of nanotechnology.

II. Background Information

1. Nanotechnology and its Potential Benefits

Nanotechnology has been touted as the next technological revolution. Nanotechnology deals with working and manufacturing on an extremely small scale. Current technology is still mostly at the micro level and not the **nanoscale**, but technological advancements have resulted in techniques that make working and manufacturing on a very small scale possible.

Areas that could be affected by the potential applications of nanotechnology include the creation of new materials by nanoengineering, the medical field, military, and defense. An example of nanotechnology applications in the medical field is the theoretical design of an artificial red blood cell. Robert Freitas, Jr., is a researcher who has designed a **nanorobot** that is a mechanical cell whose function is to mimic the actions of a natural red blood cell. Freitas calls his mechanical cell a **respirocyte**. Its size is roughly 1 micron in diameter, which means it is small enough to float along in the bloodstream. It is made of approximately 18 billion atoms that are mostly carbon. The respirocyte is essentially a tiny pressure tank that could be pumped full of oxygen (O_2) and carbon dioxide (CO_2) molecules.

The advantage of Freitas' respirocyte is that it could deliver 236 times more oxygen per unit volume than a natural red blood cell. His design includes an onboard computer that would detect gas concentrations and serve as a control for the loading and unloading of the gases (O_2 and CO_2) from the tanks. Potential military and defense applications range from nanosensors integrated into uniforms that are capable of detecting exposure to biotoxins to using **carbon nanotubes** to develop military equipment that is stronger, faster, and lighter. These types of possible applications have excited scientists and engineers working in nanotechnology. See the Resource Index for Web sites with more information.

2. Nanotechnology and Associated Risks

For scientists studying nanotechnology, defining risk/benefit trade-offs generally means weighing the risks and benefits of **nanoparticles**, fullerenes, nanotubes, and nanowires and comparing them to those associated with materials that are currently in use. For example, the application of **nanoshells** to treat cancer has become the source of concern for some scientists.

Nanoshells, made of silicate or silver core nanoparticles surrounded by a gold coating, have unique optical properties. It is hoped that chemically modified nanoshells could identify, bind to, and selectively destroy cancer cells. The results could be a significant reduction of chemotherapy side-effects and a higher survival rate due to early detection of cancer cells. But some concerns about the potential hazards of nanoparticles and nanotubes for human health and the environment have been raised. These concerns center around the extremely small size of these nanoshells. There is worry that nanoshells may be able to penetrate the skin and possibly even elude the immune system to reach the brain. From an environmental standpoint, issues such as the pace and strength with which nanomaterials may bind to organisms and nonliving species in water, soil, and air, as well as their stability over time and potential bioaccumulation in the food chain, are being discussed.

There are numerous examples of human health risks and environmental hazards as a result of using certain types of substances. **Asbestos** represents a good example. Asbestos is used in insulation, flooring, and roofing. Asbestos materials consist of tiny fibrous minerals found naturally throughout much of the earth. It is reported that roughly two-thirds of the rocks in the earth's crust contain asbestos fibers. These fibers are also found in drinking water, especially in areas near asbestos mines, such as the ones found in Quebec. Asbestos is not harmful when ingested through drinking water in certain amounts, but asbestos is considered to pose serious health hazards when inhaled. It has been linked to lung cancer, mesothelioma, and asbestosis. Asbestos is an example of a natural material that had unintended health hazards when engineered. What unintended health hazards could nanoengineered materials have? See the Resource Index for Web sites that can give you more details.

3. Societal and Ethical Concerns

Nanotechnology receives a great deal of monetary support from government initiatives in areas of defense, medicine, engineering, and education. Nanotechnology certainly has the momentum and financial backing necessary for continued growth and exploration. However, some people are concerned about nanotechnology's impacts on society. It is important to consider the ethical and societal concerns.

One concern deals with issues of **privacy invasion**. With technology getting smaller and smaller, some speculate that devices will be created to monitor people without them knowing they are under surveillance. For military and security purposes, this would be a good technological advancement. However, if the technology were used in the public sector, this would result in people being under surveillance that is virtually undetectable. Remember that a strand of human hair is roughly 50,000 nanometers in diameter. Therefore, devices that truly are nanoscale could be used without anyone's awareness. Taken further, some people speculate that pills (similar to the silicon boxes used with diabetic rats) or implants with monitoring technology could be administered to unknowing individuals and be used to detect their every move. See the Resource Index for Web sites that can give you more details.

Nanotechnology represents a great potential to revolutionize medicine and defense. It is pursued primarily for these reasons. However, it is of utmost importance that we consider all ethical and societal issues regarding nanotechnology so that proper legislation and governance can be implemented to ensure its proper use.

4. Need for Guidelines

The ultimate goal for making advancements in technology is to help humans work and perform activities more efficiently. Unfortunately, with good also comes bad. The potential benefits of nanotechnology are great. However, there are legitimate concerns regarding the potential harmful effects of misuses of nanotechnology. The goal of exploring both the benefits and risks of nanotechnology is to provide an awareness of the potential ethical and societal issues that need to be addressed. To minimize the potential harmful impacts of nanotechnology, it is important to develop certain legislative measures and systems of governance. See the Resource Index for more information.

5. Employment in Nanotechnology

Another benefit of nanotechnology is in the form of the enormous employment that this field's explosive growth will require. Dr. Mikhail Roco, the senior adviser for Nanotechnology, National Science Foundation, and chair of the subcommittee on Nanoscale Science, Engineering and Technology, estimates that *2 million nanotechnology workers will be needed* in the next 10 to 15 years.

As more knowledge is gained and more advancements are made, more individuals are deciding to pursue careers in nanotechnology. Future applications of nanotechnology encompass many different disciplines and therefore it has the potential to significantly stimulate the job market and the economy in these disciplines.

Key Terms

nanotechnology	respirocyte	nanoshells
nanoscale	carbon nanotubes	asbestos
nanorobot	nanoparticles	privacy invasion

III. Student Materials

Use Project 4 Design Brief: How Does the Public Feel about Nanotechnology? and Background Information: How to Design and Conduct a Survey to complete the projects.

UNIT 4: PROJECT 4   DESIGN BRIEF: HOW DOES THE PUBLIC FEEL ABOUT NANOTECHNOLOGY?

You are the deputy mayor of Microville, USA. An industry leader for nanotechnologies has proposed to build a large research and manufacturing facility for nanotechnology-related products in your town. As the deputy mayor, your job is to determine the potential impact (both positive and negative) nanotechnology will have on the town and the public in general. In doing some preliminary research, you have discovered that, in addition to widespread benefits, there are many societal and ethical concerns surrounding the advancement of nanotechnology.

You have decided to hold a town meeting that will bring together proponents and opponents of nanotechnology. To be able to effectively respond to both sides of the debate, you will conduct comprehensive research on nanotechnology. This research will include a summary of the risks and benefits of nanotechnology, a survey designed to assess the public's knowledge of and attitude toward the technology, and a final report/presentation that will be given at the town meeting. The purpose of your research and presentation is to describe the potential risks and benefits of nanotechnology as well as describe the general public's knowledge of and attitude toward this emerging technology.

Design Constraints: Things You Must Do

Below are constraints, or guidelines, to follow during this design brief:

1. Research and compose a brief description of the potential benefits (manufacturing, health, and defense) of nanotechnology.

2. Research and describe the key issues that represent the health, environmental, ethical and societal concerns related to the potential development and use of nanotechnology.

3. Using the information you gather during steps 1 and 2, design and conduct a survey that assesses the general public's knowledge of and attitudes toward nanotechnology. Follow the guidelines in the "Background Information: How to Design and Conduct A Survey" section on p. 158.

4. It is suggested that your survey be in the form of personal interviews. You should have a sample of *at least* 10 people. These people can be classmates, friends, relatives, or neighbors.

5. Be sure that the survey you design includes:

   a. A place to gather demographic information (e.g., age, gender, level of education, and occupation) from the interviewees.

   b. Questions that will give you the information you are seeking. See the Resource Index for a Web site with survey design tips.

   c. Allows you to collect data that can be easily recorded, organized, and analyzed.

6. Once the surveys have been conducted, gather, organize, and analyze your data. You may use Microsoft Office Excel® to help you.

7. Your final presentation should include:

   a. An Introduction/Overview.

   b. A Methods section in which you describe how the survey was designed and conducted. Be sure to include a description of your sample.

   c. A Results section in which you present what you found. Be sure to include graphical representations of your data.

   d. A Discussion section in which you describe the limitations of your study, ideas for future studies, and conclusions.

Background Information: How to Design and Conduct a Survey

Surveys are used in many fields to gather data on the knowledge, attitude, and behavior of the general public or specific populations. This information may be used to design political campaigns, create marketing strategies for new products, and many other purposes. Below is an outline that will help you as you design and carry out your survey on the public's knowledge of and attitudes towards nanotechnology.

a. The Purpose of Sampling

In most cases it is impossible to study an entire population, therefore we study a sample population (a small group of individuals often randomly selected from a population) and apply the conclusions from the sample study to the whole population. This is often accomplished through survey research.

b. Surveying

Typically there are three kinds of information you can gather through surveying:

- Descriptions about the respondent and their background
- Behaviors
- Preferences or opinions

Generally speaking, there are several ways to conduct a survey:

- Mail-out
- In person (interview)
- Telephone

c. Steps in Conducting Survey

1. Identification of the purpose and focus of the survey
2. Determine population to be surveyed
3. Design survey instrument
4. Pretest of survey instrument to be sure you are getting the kind of information you are seeking
5. Implementation of survey
6. Gather and organize the data (a computer may help here)
7. Analyze the data collect and write a report detailing what you found

d. Types of Bias

Be aware that as with any type of research, surveys are subject to a certain degree of bias. There are several potential sources of bias listed below:

Response bias—when a respondent tries to tailor his response to what he thinks the interviewer wants
Wording—using leading statements often affects the answers to questions
Choices—the choices or ranges given for questions can also change the perceived conclusions

PROJECT 5
Forces and Effects at the Nanoscale

I. Project Lesson Plan

1. Project Description

During this project, you will explore ways in which the characteristics and properties of matter differ when one moves from the macroscale to the nanoscale. You will actively investigate several of the key forces and effects (e.g., surface effects and self-assembly) at this small scale.

2. Learning Objectives

- You will describe some of the properties and characteristics of matter at the macroscale and nanoscale.

- You will compare and contrast the forces and effects that dominate at the macroscale and nanoscale.

- You will be able to apply knowledge of nanoscale forces and effects to the construction of nanoscale components.

II. Background Information

1. Changes at the Nanoscale

In addition to changes in optical properties (e.g., color, luster, and fluorescence), the electrical (e.g., conductivity) and mechanical (e.g., strength and flexibility) properties of nanosized matter may change. But what is the reason for such changes? There are four main factors or forces at play that are different when we consider nanosized particles of substances. First, due to the extremely small mass of these particles, *gravitational forces* are almost nonexistent. Instead *electromagnetic forces* dominate and ultimately determine the behavior of atoms (and molecules). Second, at nanoscale sizes, we need to use *quantum mechanics* to describe the particle motion and energy transfer instead of the classical mechanics descriptions. Third, nanosized particles have a very large *surface area-to-volume ratio*. Fourth, at this size, the influences of random thermal molecular motion play a much greater role than they do at the macroscale.

The Design Brief activity in Project 2 provided a demonstration of gravitational versus electromagnetic forces. This project looks at the importance of surface area to volume ratio to understanding this effect. An additional activity gives a demonstration of how surface characteristics can be used in the self-assembly of nanosized objects.

2. The Importance of Surface Effects

Nanoscale objects have a far greater amount of surface area than volume, so surface effects (e.g., surface area, gravitational forces, and surface forces) are far more significant in general. Keep in mind that the smaller something is, the larger its surface area is compared to its volume. This high surface-to-volume ratio is an important characteristic of nanoparticles. Figure 4-12 illustrates this area. It shows cubes of varying sizes and the impact that changes in size have on various dimensions. Each time the length doubles, the volume increases by eight-fold. Notice that the smaller the cube, the larger the surface area-to-volume ratio.

Now just imagine how great the surface area to volume ratio would be for something as small as nanoscale particles. The huge ratio of surface area to volume makes interactions between the surfaces of particles and their substrate extremely important to nanoscientists. They know that most reactions occur at the interface of two substances, so an increased surface area means increased reactivity and that is why nanosized groups of particles have the potential to make great catalysts.

Why doesn't a sugar cube stick to me as well as powdered sugar does? The ability of a sugar cube to stick depends on a balance of size-dependent forces (Figure 4-13). That is, as the sugar cube gets smaller, adhesive forces increasingly dominate the gravity forces, and thus the smaller sugar cubes stick to us better than the large ones. The relationships among the sizes of the sugar and the strength of the forces acting upon them can be shown as in Figure 4-13.

This principle is explored in the design briefs for this project (see pages 164–166).

3. Working at the Nanoscale

Working and building at the nanoscale isn't quite as straightforward as simply making your tools smaller and using powerful microscopes. Objects at this scale start to become very "sticky." Nanoparticles are attracted to each other via electrostatic forces, and this effect makes it hard to handle and move things that are

Figure 4-12 Example of How an Object's Size Impacts Its Surface Area-to-Volume Ratio

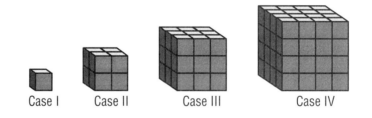

Parameters	Case I	Case II	Case III	Case IV
Length (L)	1	2	3	4
Face Area (L^2)	1	4	9	16
Volume (L^3)	1	8	27	64
Surface Area ($L^2 \times 6$ faces)	6	24	54	96
Area/Volume ratio	6	3	2	1.5

Figure 4-13 Adhesive and Gravitational Force The ratio of surface to volume can determine whether adhesion or gravity will dominate.

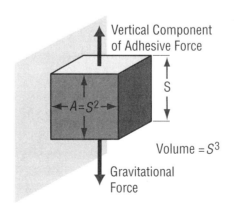

Figure 4-14 IBM Logo Assembled from Individual Xenon Atoms Arranged on a Nickel Surface

extremely small. Nanoscientists have used several techniques that allow them to manipulate atoms and molecules in an attempt to learn how to build things at the nanoscale (nanofabrication).

One way to do this is to use *atom-by-atom assembly*, which is similar to brick-laying. Atoms are moved into place one at a time using special tools such as the scanning tunneling microscope (STM) or an atomic force microscope (AFM). Both of these special pieces of equipment have been discussed in more detail in Project 3. As you might guess, building structures one atom at a time is quite expensive and tedious, but possible. Figure 4-14 shows 35 xenon atoms positioned to create a nanoscale structure in the form of the IBM logo.

Another approach to nanofabrication is **self-assembly**. This typically involves setting up an environment in which structures assemble automatically. Examples include chemical vapor deposition and the patterned growth of nanotubes. Fortunately, there exist numerous examples of self-assembly mechanisms in nature and our ability to create nanostructures improves as we gain more sophisticated understandings of biological self-assembly.

4. Biological Self-assembly and Naturally Occurring Molecular Machines

Molecules try to minimize their energy levels by aligning themselves in particular positions. If bonding to an adjacent molecule allows for a lower energy state, then the bonding will occur. Self-assembly in nature is older than life itself; in fact, all living organisms, from the simplest single-cell species to humans, depend on some form of molecular self-assembly. Protein folding, nucleic acid assembly, cell membranes, ribosomes, and the capsides of viruses are but a few representative examples of biological self-assembly. Salt will assemble to form crystals. Salt crystals form as the salt molecules arrange themselves while the water evaporates. The bonds between the salt molecules are strong enough to squeeze out the water and arrange themselves to form a crystal. The different geometries of the salt molecules affect the shape of the salt crystals, so the nanoscale geometry affects the macroscale appearance of the crystal. Figure 4-15 shows tiny spheres of polystyrene self-assembling.

Soap bubbles also self-assemble. The soap molecules form two layers that sandwich a layer of water in between (Figure 4-16). This is because the soap molecules have one end that likes water and one that does not. So the end that does not like water is on the outside and the other end that likes the water is on the inside. The soap forms a monolayer on the inside and a monolayer on the outside of the water. Each layer of soap is a self-assembled monolayer—a single layer of molecules oriented in one direction. Interestingly, the structure described previously is remarkably similar to the way in which cell membranes are constructed. For more information on cell membrane self-assembly and the structure of biological membranes, see the Resource Index.

Figure 4-15 An Image of Polystyrene Spheres Self-assembling

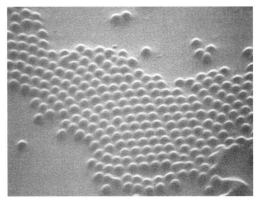

(Courtesy of Nanoptek Corporation)

Figure 4-16 Soap Bubble, Water Soap molecules form a self-assembled monolayer on either side of a layer of water.

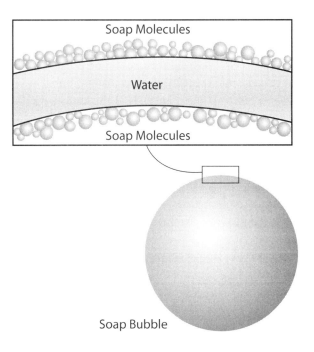

Designing and building **nanostructures** and **nanomachines** by arranging and manipulating atoms and molecules may seem impossible, but that is the level at which nanoscientists operate. Once again researchers in this field can thank nature for providing them with terrific examples of how to build at the nanoscale. Unlike machines that are manufactured, molecular machines are naturally occurring devices that perform specific functions for living things (Figure 4-17).

Examples of molecular machines can be found within our bodies. **Ribosomes**, for example, are found within our cells and are responsible for building proteins through protein synthesis. The way in which ribosomes build proteins is by stringing together amino acids, which are the building blocks of proteins, in a precise sequence. The instructions ribosomes follow to build proteins are provided by messenger RNA (mRNA). Stringing together these amino acids is an example of building at the molecular level. Ribosomes are machinelike in that they are responsible for the construction of the protein. Ribosomes read the mRNA one codon at a time.

For every codon in the mRNA, there is a tRNA (transfer RNA) molecule that fits the codon exactly. The tRNA acts like a delivery truck that carries the

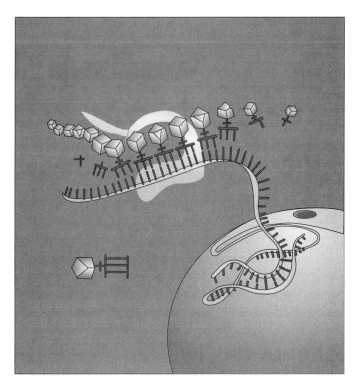

(Courtesy of Canadian Museum of Nature)

Figure 4-17 Ribosome Constructing a Protein
Ribosomes act like machines by stringing together amino acids in a precise sequence to build a protein.

proper amino acid to the ribosome when the codons call for them. The ribosome essentially moves along the mRNA, decodes the instructions, and then adds a new amino acid to the chain in order to create the protein. Figure 4-17 is an illustration of a ribosome constructing a protein.

Ribosomes, which are approximately 20–30 nanometers in diameter, are naturally occurring machines in that they act as both a decoding device as well as a construction device. Ribosomes are just one example of a molecular machine that occurs naturally in our bodies. For more information about ribosomes, see the Resource Index.

With advances in our understanding of nanotechnology, researchers hope to use the bottom-up approach to build manufactured molecular machines. Just as ribosomes move along mRNA building a protein, imagine a manufactured nanoscale device that could build other nanoscale devices. Researchers call these nanorobots and nanoprobes. They hope to use these molecular assemblers to construct additional assemblers and ultimately to create molecular machines that have been designed to atomic specifications.

Key Terms

self-assembly   nanomachine   rate of reaction
nanostructure   ribosomes

III. Student Materials

Use Project 5 Design Brief 1: Surface Area-to-Volume Ratio and Project 5 Design Brief 2: Self-Assembly to complete the project.

UNIT 4: PROJECT 5 DESIGN BRIEF 1: SURFACE AREA-TO-VOLUME RATIO

Introduction

The overarching goal of nanoscience is the ability to manipulate materials at the atomic and molecular level; however, today's manufacturing technology is too large and bulky to manufacture things at this scale. This difficulty is due mainly to the changes in the properties and characteristics of materials with this extremely small size. This series of activities is designed to help you better understand the forces and effects that dominate at the nanoscale and ultimately govern the way in which nanoscientists work at the nanoscale.

Part A

The relative strength of forces acting on an object can depend on the size of the object. Nanoscale objects have a far greater amount of surface area than volume, so surface effects (e.g., surface area, gravitational forces, and surface forces) are far more significant in general. In this activity, you will examine the impact that the surface area-to-volume ratio of an object has on its behavior.

Student Materials

Granulated sugar
Powdered sugar
Sugar cubes
Large Ziploc® plastic bags (gallon size)
Paper towels

Procedures

1. Have one member fill one large bag about two-thirds full with sugar cubes; one bag with granulated sugar about two-thirds full; and a third bag with the same amount of powdered sugar.

2. Have one member hold up a bag while another team member carefully dips his or her hand into the bag. Make and record observations. The team member who dipped his or her hand into the bag should wipe his or her hand clean with a dry paper towel. Repeat this procedure for all three bags of sugar.

Questions to Answer

1. Did each substance stick to the hand?

2. How much of the substance stuck?

3. What did it feel like?

4. Could the sugar be easily shaken off?

Part B

Which size substance dissolves the fastest? Another example showing the importance of surface area-to-volume ratio of matter is its **rate of reaction**. Since reactions occur at the interface of two substances, when a large percentage of the particles is on the surface, we get maximum exposed surface area, which means maximum reactivity…so *nanosized* particles may make great catalysts. The purpose of the activity is to compare the effects of varying the surface area to the volume ratio for two samples of the same mass and type of substance on the rate of dissolving in water.

Student Materials

Granulated sugar
Powdered sugar

Sugar cubes
Erlenmeyer flasks (3 per group)
Weighing paper
Electronic balance
Graduated cylinder
Water
Grease marker

Procedure

1. Using a grease marker, label one Erlenmeyer flask #1, one #2, and the other #3.

2. Measure and record the mass of one cube of sugar. Use a piece of weighing paper. Put the sugar cube into flask #1.

3. Measure and record an equal mass of granulated sugar and put the granulated sugar into flask #2.

4. Measure and record an equal mass of powdered sugar and put it into flask #3.

5. Using your graduated cylinder, carefully measure and add 100 milliliters of tap water to flask #1.

6. Have one group member hold flask #1 at the neck and gently swirl the flask exactly 60 seconds.

7. Observe and record the relative amount of the sugar cube that has dissolved in the flask. You can do this visually, while describing your observations, but you may want to taste a sample of the water and record how much sugar you can taste.

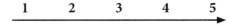

1 = no sugar taste
2 = weak sugar taste
3 = mild sugar taste
4 = moderate sugar taste
5 = strong sugar taste

8. Repeat steps 5, 6, and 7 for flasks #2 and #3.

Questions to Answer

1. Describe the variation that exists in the relative amount of sugar that dissolves in 60 seconds.

2. Describe the visual appearance and taste of the three solutions.

UNIT 4: PROJECT 5   DESIGN BRIEF 2: SELF-ASSEMBLY

Introduction

One of the proposed methods to make nanosized objects and machines is to use nature's own tendency to self-assemble. Self-assembly in nature is older than life itself; in fact all living organisms, from the simplest single-cell species to humans, depend on some form of molecular self-assembly. Protein folding, nucleic acid assembly, cell membranes, ribosomes, and the capsides of viruses are but a few representative examples of biological self-assembly. For example, salt will assemble to form crystals. Salt crystals form as the salt molecules arrange themselves while the water evaporates. The bonds between the salt molecules are strong enough to squeeze out the water and arrange themselves to form a crystal. But what does self-assembly look like? How do we know when it has taken place? Luckily the process of self-assembly can be easily demonstrated using soap bubbles, which will self-assemble when one is popped.

Student Materials

Bubble solution
Large beaker
Toothpicks
Hand lenses

Procedure

1. Obtain the previous list of materials. Have one team member pour about 10 milliliters of the bubble solution into a clean beaker.

2. Cover the top of the beaker with your hand and shake vigorously. This should produce a solution with lots of bubbles on the top. Be careful not to spill the solution or to drop the beaker!

3. Take the toothpick and pop one of the bubbles. Notice how the arrangement of bubbles changes.

4. Repeat this procedure several times. Clean the beaker between repeats.

Design Constraints: Things You Must Do

Below are constraints, or guidelines, to follow during this design brief:

1. Work in pairs to conduct these experiments.

2. Follow all procedures and instructions carefully.

3. Record all observations and results as you conduct the experiments.

4. You will create a lab report that details the experiments and results from each part of the Design Brief. The completed lab report should include:

   - An Introduction section explaining the purpose of the laboratory experiment.

   - A detailed description of what was done in a Methods section.

   - A Results section that includes sketches and notes of observations that were made.

   - The Results section should contain a table(s) and graph(s) of the data.

   - A Discussion section that explains the results and explains any problems encountered that might have affected the results.

   - A Conclusion section that summarizes the experiments and links back to the nanoscale.

PROJECT 6
Applications in the Field of Nanotechnology

I. Project Lesson Plan

1. Project Description

In this project, you will extend your knowledge of cutting edge applications of nanotechnology. Using the background on nanotechnology provided in this project and in Project 3, you will examine some of the current and future applications of nanotechnology in the areas of manufacturing, medicine, and defense. The project culminates in a design brief in which you assume the role of a design engineer for a private nanotechnology firm whose task is the development of a presentation at a local town meeting.

2. Learning Objectives

- You should be able to understand the relationship between nanoscience and nanotechnology.
- You will detail the current and future applications of nanotechnology.
- You will develop a presentation design to inform the general public about the emerging field of nanotechnology.
- You will develop a prototype of a nanomachine designed to perform a specific beneficial task.

II. Background Information

1. The Connection between Nanoscience and Nanotechnology

Studying the fundamental physical, chemical, and biological properties of objects such as atoms, molecules, and structures at the nanoscale is what nanoscientists focus on. Applying that knowledge to advance technology is what nanotechnologists focus on. Nanotechnology, therefore, is the application of nanoscience at the atomic or molecular scale of approximately 1–100 nanometers. It consists of the tools and techniques used to produce nanoscale objects and devices.

For example, a nanoscientist studies the fundamental or underlying properties of **carbon nanotubes**. Nanotechnologists design and develop uses for carbon nanotubes, such as using them to build stronger and faster military equipment or medical devices.

It is difficult to imagine having the ability to manufacture structures atom by atom, but that is exactly what researchers in nanotechnology hope to do. Researchers in the field of nanotechnology come from a variety of disciplines such as materials science, electronic engineering, mechanical engineering, and medicine. There are two approaches regarding how to design and develop in nanotechnology: *top-down* and *bottom-up*. Researchers using the top-down approach are developing machining and etching techniques to create nanoscale structures. This is analogous to taking a larger bulk substance and whittling away the substance until you reach the nanoscale. In contrast to this approach, researchers who adhere to the bottom-up approach are investigating ways to build and manufacture structures, such as molecular machines, molecule by molecule or even at the level of atom by atom. It may seem like it would be impossible to build using a bottom-up approach, but there are examples of naturally occurring molecular machines that do indeed build at the molecular level.

2. Proposed Applications of Nanotechnology

Although much is already known about objects at the nanoscale, current technology lacks the tools to enable engineers to build and manufacture at atomic specifications. Much of what is discussed in nanotechnology is theoretical designs and potential applications. The two areas where research and advances in nanotechnology have the greatest potential impact are medicine and military defense.

a. Potential Medical Applications of Nanotechnology

Researchers hope to one day design *nanoprobes* that could be used to detect and destroy viruses within the body (Figure 4-18). Other researchers have proposed nanoprobes that could emit small amounts of radiation that could kill bacterial and viral organisms.

Researcher Robert Freitas, Jr., has designed a **respirocyte**, a mechanical cell that mimics the actions of a natural red blood cell. Its size is roughly 1 micron in diameter, which means it is small enough to float along in the bloodstream. It is made of approximately 18 billion atoms that are mostly carbon.

The respirocyte is essentially a tiny pressure tank that could be pumped full of oxygen (O_2) and carbon dioxide (CO_2) molecules. See Figure 4-19 for an illustration of the respirocyte. The advantage of the respirocyte is that it could deliver 236 times more oxygen per unit volume than a natural red blood cell. Freitas' design includes an onboard computer that would detect gas concentrations and serve as a control for the loading and unloading of gases from the tanks. It is important to realize that these types of designs are only theoretical at this point and are not available.

Figure 4-18 Conceptual Representation of a Nanoprobe Used in Medicine Researchers hope to design nanoprobes to detect and destroy viruses within the body.

(Courtesy of Coneyl Jay/Photo Researchers, Inc.
© 2006 All rights reserved.)

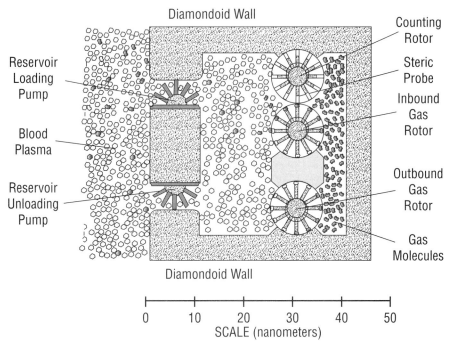

Counting Rotor

Steric Probe

Inbound Gas Rotor

Outbound Gas Rotor

Gas Molecules

Diamondoid Wall

Reservoir Loading Pump

Blood Plasma

Reservoir Unloading Pump

Diamondoid Wall

SCALE (nanometers)

Figure 4-19 Theoretical Design of a Chemical Concentration Nonsensor Embedded in the Hull of an Artificial Nanorobotic Red Blood Cell The artificial red blood cell, or respirocyte, is essentially a tiny pressure tank that could be pumped full of oxygen (O_2) and carbon dioxide (CO_2) molecules. It could deliver 236 times more oxygen per unit volume than a natural red blood cell.

(Copyright 1996, Robert A. Freitas Jr.)

b. Potential Defense Applications of Nanotechnology

Nanotechnology is a multidisciplinary subject, so it is important to maintain coordinated efforts. The National Nanotechnology Initiative (NNI) is aimed at coordinating the efforts of the many federal agencies involved in nanotechnology. The Department of Defense, National Science Foundation, Department of Homeland Security, and Department of Energy are just a few of the agencies involved in the NNI.

In 2005, President Bush asked for $982 million for the National Nanotechnology Initiative. In 2009, the federal budget provided $1.5 billion in funding to the National Nanotechnology Initiative. This funding is then distributed by the NNI to other government agencies, including the National Science Foundation and the Department of Defense. See Table 4-2 for the allocation of research dollars to these two agencies.

In the area of defense, researchers hope to use nanotechnology to improve the equipment used in the military and to detect threats such as biological substances. One example of a potential use of nanotechnology is in the manufacturing of soldiers' uniforms and equipment. *Nanosensors* could be used to detect exposure to airborne chemicals or other toxins. Fabric could also be reinforced with *nanoparticles*, leading to lightweight helmets and other equipment. It has even been suggested that a material could be manufactured that could repair itself when ripped.

There are many examples of proposed uses and applications for nanotechnology. However, the technology is not yet advanced enough to begin mass producing and manufacturing products at the nanoscale level. For this reason, it is important to realize that much of what is discussed regarding nanotechnology are proposals that represent theoretical applications and designs.

Table 4-2 Research dollars (in Millions)

Year	Department of Defense	National Science Foundation
2007 Actual	$450	$389
2008 Estimated	$487	$389
2009 Proposed	$431	$397

3. Current Applications of Nanotechnology

The current state of nanotechnology is far from the ability to produce nanoprobes and nanosensors but some products do currently utilize nanotechnology. It should be noted that this does not necessarily mean these products were nanoengineered at the atomic level. Certain nanomaterials can be purchased as a liquid dispersion or as a dry powder and can then be incorporated into an existing product in hopes of improving the functionality of the product.

According to the NNI, current products that benefit from utilizing nanomaterials are bumpers on cars, coatings to protect against corrosion, coatings to reduce glare for eyeglasses, sunscreens, tennis balls, and tennis racquets. Some clothing items utilize nanomaterials to make them stain-resistant. See the Resource Index for more information.

Key Terms

carbon nanotubes   respirocyte

III. Student Materials

Use Project 6 Design Brief: Why Your Town Needs Nanotechnology to complete the project.

UNIT 4: PROJECT 6   DESIGN BRIEF: WHY YOUR TOWN NEEDS NANOTECHNOLOGY

You are employed by The Reed Park Corporation, which is a large company that designs cutting-edge nanotechnology applications. Your company is trying to build a new facility in the town of Boiling Springs. While the new facility will provide many jobs for the town's people, the company has met some resistance. This resistance is due primarily to the fact that many of the town's residents don't know much about nanotechnology. They are also not aware of how the company will impact their town.

As a design engineer in Reed Park's technology marketing department, your responsibility is to develop a presentation that can be used to help educate the town about what nanotechnology is and the types of applications scientists working in this field are pursuing. Your supervisor has asked you to give this presentation at an upcoming local town meeting in Boiling Springs.

As part of this presentation, you are to showcase an original design for a nano-machine made to execute a task that will help the town of Boiling Springs. This will certainly prove that you are an expert in the field and will help convince the local townspeople that nanotechnology will be beneficial to their community.

Design Constraints: Things You Must Do

Your presentation should be 10 minutes in length and should include a:

- Brief introduction of the multidisciplinary nature of the field of nanotechnology.

- Discussion of why it is important for the town's residents to have basic knowledge of the field of nanotechnology.

- Description of the field of nanotechnology, providing examples of the tools of the trade and current applications. Include images of your examples.

- Discussion of the potential of this emerging field. Include examples from manufacturing, medicine, and defense. Detail the amount of money that is being directed to research in this area.

- Design of a prototype of your nanomachine intended to execute a particular task that will help the community. Draw your machine, labeling its major parts. Be sure to describe what job your nanomachine will do and give it a name.

- The final slide of your presentation should include the following:

  - The URLs for Web sites that provide meeting attendees with additional information about nanotechnology

  - The citations for where you found the images you used in your presentation

UNIT 5
Biometrics

Unit Overview

I. Introduction

Biometrics is the science and technology of collecting and analyzing data about physical or behavioral individual traits. This data is used for identifying or verifying the identity of a person. While biometric techniques date back to ancient civilizations, modern technologies that are able to capture unique physical characteristics of individuals have exploded in recent years. Biometric traits recorded and analyzed by these new technologies include **fingerprints**, facial patterns, iris or retinal patterns, voice patterns, hand geometries, and vascular patterns. As more of our personal information is held in electronic form, it has become increasingly important to be able to link this information with individuals.

II. Unit Learning Goals

- You will learn about the foundations and history of biometrics technology.
- You will learn about fingerprint classification and the technologies used to collect, enhance, and analyze them.
- You will research the advantages and disadvantages of biometrics with respect to identity theft and privacy.
- You will learn about the application of 2D and 3D visualization software in crime scene recreation and how biometric evidence is integrated with this information.

- You will gain in-depth knowledge about one area of biometrics through the design of a biometric device.

III. Projects

Introductory Level Projects

PROJECT 1: Biometric Tools

You will learn about the history and underlying technology of numerous biometric tools currently available as well as those under development. This project will provide a foundation for the study of biometrics and investigate the history, products, and uses associated with biometric technology.

PROJECT 2: Collecting and Analyzing Fingerprints

You will explore the most common biometric measure—the fingerprint. You will learn how to collect, analyze, and match fingerprints using existing classification systems.

PROJECT 3: Processing Fingerprints

This project introduces you to fingerprinting applications and processing techniques associated with various fingerprint biometric devices. As part of your design brief, you will match a crime scene fingerprint with a suspect's fingerprint by using image processing and highlighting techniques.

PROJECT 4: Identity Theft and Privacy

You will discuss the implications of using biometric devices in our society, including issues of identity theft and privacy. In the design brief, you will take a stance for or against the use of biometrics as it relates to identity theft and personal privacy.

Intermediate Level Projects

PROJECT 5: Crime Scene Visualization

You will use 2D and 3D visualization software tools to recreate a crime scene. This scene will be used to analyze the current knowledge and evidence pertaining to the crime, including positions of individuals over time, line of sight, and proximity to evidence. These visualization can be crucial for locating biometric-based evidence collected at the crime scene in space and time and for re-creating its role in the crime.

PROJECT 6: Biometric Device Design

You will take what you have previously learned about biometric concepts and techniques and apply it to the design of a biometric device. This device will both represent the science of biometrics and meet a known need for biometric measurement and analysis.

Advanced Project

You will complete an independent project through the use of visualization tools by researching a new topic dealing with bioprocessing or by expanding on topics covered in this unit. The objective of the advanced level is for you to further your skills in integrating research, problem solving through the design brief approach, and presentation. It is up to the teacher to work with you to negotiate the topic, time allocated to the project, and design constraints.

IV. Unit Resources

The Resource Index contains a listing of all resources associated with the unit. Included are relevant Web site links, books, and other publications. The Glossary provides definitions for all Key Terms listed in each project.

PROJECT 1
Biometric Tools

I. Project Lesson Plan

1. Project Description

You will learn about the history and underlying technology of numerous biometric tools currently available as well as those under development. This project will provide a foundation for the study of biometrics and investigate the history, products, and uses associated with biometric technology.

2. Learning Objectives

- You will learn about the foundations and history of biometrics technology through class discussion.

- You will learn about current and future biometric tools, their advantages and disadvantages, and applications by researching individually or with groups.

- You will work individually or with a group of students to become experts in a biometric tool of your choice by conducting research that will eventually lead to the development of a class Web site on biometrics.

- You will learn how to use Web development software by developing a Web site about biometric technology.

- You will investigate design issues associated with biometric technology by developing a future application for a chosen biometric.

II. Background Information

1. What Is Biometrics?

Biometrics is the science and technology of collecting and analyzing data about physical or behavioral individual traits used for identifying or verifying the identity of a person. Biometrics literally means to measure life, derived from the Greek words *bio* (life) and *metric* (to measure). Biometric traits include **fingerprints**, facial patterns, iris or retinal patterns, voice patterns, hand geometries, and vascular patterns.

Biometric features have been used for thousands of years to identify people based on facial features, body measurements, signatures, and fingerprints. Current biometrics involves the automation of identification processes.

To be considered a **biometric** trait, a physical or behavioral characteristic must be:

- Universal: present in every human being
- Unique: pattern or form is exclusive to a specific individual
- Permanent: stable over time
- Measurable: collectable and quantifiable

Biometrics falls into two main categories:

1. **Identification**: Who is it?

2. **Verification** or **authentication**: How can you be sure?

Identification is the more difficult task of the two categories since it involves comparing a given trait, such as a fingerprint, to a database or library of characteristics for a large population. It is most often used by law enforcement agencies.

On the other hand, verification, also known as authentication, tries to make a match between a biometric trait and another piece of information. Most frequently, a person provides a piece of identification or unique information and then is prompted to provide a biometric trait. Other times, a template of a person's biometric trait already exists. When seeking verification, the person needs to provide only the trait for comparison. In both cases, a one-to-one match must be made between the identification or unique information and the biometric trait to authenticate an individual.

Authentication is most often used to gain access or entry to a specific site or service. Because biometric systems cannot be lost, easily stolen, or duplicated, they are often considered to be more secure than traditional forms of identification, such as personal identification numbers (PINs), cards, passwords, or keys.

2. History of Biometrics

The practice of biometrics, although not by that name, dates back to the ancient Egyptian and Asian civilizations, where traders and artisans were identified by fingerprints on their merchandise or by physical characteristics such as eye color, complexion, and height. The earliest records of biometrics are a 10,000-year-old partial handprint and 4,000-year-old handprints and footprints found in Egypt. Most likely, these prints were used to establish ownership of belongings. Fingerprints were also used in China in the third century B.C. on legal documents as well as in court (Figure 5-1).

Fingerprints are probably the oldest example of a biometric still in use today. Despite the use of fingerprints by ancient civilizations, not much was known or documented about the characteristics of fingerprints until the seventeenth century, when Dr. Nehemiah Grew, an Englishman, wrote about the anatomy of fingerprints. Around the same time, Italian professor Marcello Malpighi, among others, identified anatomical characteristics unique to the skin of fingers and palms.

However, significant advances in the science of fingerprinting did not come until the nineteenth and twentieth centuries. In 1823, Professor Johannes Evangelist Purkinje conducted in-depth studies of the skin on fingers and palms. He found and documented basic patterns, which were elaborated and refined by others.

In the late nineteenth century, Dr. Henry Faulds, a Scottish missionary doctor, provided the foundation for the classification systems used in law enforcement, which is considered one of the most influential works in fingerprinting. He made careful observations of patterns on fingertips while working at a hospital in Japan. He noted that patterns were permanent even after injury. He also devised a system for obtaining clear fingerprints using a sheet of slate or tin and printer's ink. Based on his observations, Dr. Faulds concluded that fingerprint patterns formed loops and whorls. He also concluded that if fingerprints were left on objects at a crime scene, they could lead to criminal identification.

Figure 5-1 Chinese Fingerprinting A third century B.C. fingerprint was taken from a piece of an old legal document in China.

This was an important conclusion because detectives did not rely on fingerprints for criminal identification. Instead, they relied on Bertillonage, an identification system based on the measurement of adult height, length of extremities and fingers, and head width. Bertillonage was named after Alphonse Bertillon, a clerk with the Paris police who was interested in developing easier ways of identifying convicted criminals. The system was used around the world until the beginning of the twentieth century, when it led to various errors as a result of identical measurements from different individuals. Bertillonage was eventually completely replaced by fingerprinting as a result of the work done by Dr. Faulds and others.

Another key person in the development of using fingerprints as biometrics was Sir William Herschel, an English administrator in India, who began taking palm and thumbprints of illiterate townspeople as a form of signature for legal documents. He also began fingerprinting all prisoners in jail. He observed, as did Dr. Faulds, that prints did not change over time, but he never came up with a fingerprint classification system or a method for identifying criminals.

This type of identification work came later, in 1892, when Sir Francis Galton (Figure 5-2) published a detailed study of fingerprints in his book *Finger Prints*. His work included the first attempt at an in-depth classification system for comparing fingerprints from criminals.

Juan Vucetich, an Argentinian police officer and friend of Galton's, used Galton's ideas to devise his own system for fingerprint classification. In 1892, he made the first criminal fingerprint identification. He used the new classification system to convict Francisca Rojas of murdering her two children in Buenos Aires, Argentina.

Galton's system became the foundation for modern fingerprint science. In 1900, Galton's student, Sir Edward Henry (Figure 5-3), published the first accurate and effective method of fingerprint classification, titled *Classification and Uses of Fingerprints*. His method, based on Galton's work, became known as the Henry System. The Henry System worked by placing an individual fingerprint into one of three classes: loop, arch, or whorl.

In 1901, Henry became the Assistant Commissioner of Police at Scotland Yard (the detective department of the police force in London), and founded the Central Fingerprinting Bureau. The first conviction using fingerprints in a British court took place in 1902. A decade later, the Henry system became successfully established by law enforcement agencies throughout the English-speaking world. It remains in use today.

By 1903, New York opened the first Fingerprint Bureau in its prison system. A few years later, the Federal Bureau of Investigation (FBI) used the Henry System to run background checks and clear military personnel. In 1915, the International Association for Identification was founded in the United States. And by 1924, the

Figure 5-2 (left) Sir Francis Galton Galton's book *Finger Prints* included the first classification system for fingerprints.

Figure 5-3 (right) Sir Edward Henry The Henry System placed a fingerprint into one of three classes: loop, arch, or whorl. It is still in use today.

(Mary Evans Picture Library/ Everett Collection)

(Hulton-Deutsch Collection/CORBIS)

FBI had established a formal fingerprint collection, which became the world's largest single repository for fingerprint data in 1941.

Galton's and Henry's work spurred an enormous interest in biometrics. However, their methods were manual and, therefore tedious and time consuming. One serious limitation of the Henry System is that it relies on having a full set of prints from an individual. A number of individuals experimented with a single print system, but Harry Battley and Frederick Cherrill are considered to be the inventors. Both worked at Scotland Yard in the late 1920s.

In 1930, Battley published his book *Single Finger Prints*. The Single Print system not only looked at fingerprint patterns, but also counted specific structures in those patterns (see Project 2 on page 189 for more detail on fingerprint patterns and structures). Although this system improved the Henry System, it involved extensive training and remained laborious and time consuming.

Fingerprint classification continued this way until 1960, when institutions in Paris and the United Kingdom began research on automatic fingerprint biometric systems. Also in the 1960s, the Miller brothers in New Jersey experimented with a hand reader verification device. Their ideas led to the founding of Recognition Systems, Inc., which further developed the hand reader in 1985.

Since the 1960s, studies have moved from fingerprints and handprints to other biometrics. In 1984, a company named EyeDentify developed the first retinal scan for commercial use. By the early 1990s, the biometrics industry became firmly established in science and technology. Subsequent studies have refined the accuracy of biometric identification and have led to numerous biometric tools. Biometrics technology will soon be a regular part of everyday life.

3. Biometrics Technology

A **biometrics device** analyzes a specific physical or behavioral characteristic (i.e., a biometric) that can identify an individual and compares it to a database or library containing information on that biometric. Depending on its use, the device may be either an identification or a verification system. Usually the device will look for patterns and will analyze those patterns by using computer algorithms. The device may contain an internal system of analysis or it may be linked to a larger network via a transmission channel.

The way a biometric device functions may be broken down into three steps:

1. Scanning: A scanning device records the biometric.
2. Storage: The biometric information is stored in a digital format.
3. Analysis: Software looks for specific points of comparison, that is, **match points**, between the scanned biometric and the database and makes a match.

Biometric devices are divided into two categories: those that measure physical traits and those that measure behavioral traits.

Biometric devices that measure physical traits include:

- **Fingerprint scanners**: analyze fingertip patterns
- **Face recognition systems**: measure facial characteristics
- **Hand geometry readers**: take measurements of the hand
- **Iris scanners**: analyze features of the iris (the colored ring of the eye)
- **Retinal scanners**: analyze the blood vessels in the eye

Biometric devices that use behavioral traits include:

- **Voice recognition systems**: authenticate vocal patterns
- **Signature recognition systems**: analyze signature components
- **Keystroke dynamics readers**: measure the timing and spacing of typed words

a. Fingerprint Scanners

Fingerprinting biometric devices (Figure 5-4) work by scanning a finger or fingers by using one of various technologies and creating a digital image (see Project 3 for more information on types of fingerprint scanners and minutiae). The scanner is attached to a computer that uses algorithms to look for whorls, arches, and loops as well as fine details known as **minutiae** and records these characteristics as encoded data. The data may be used to make comparisons to other fingerprint images or may be stored for future comparisons in a data-bank. By encoding the data rather than saving an image of the fingerprint, the device ensures that fingerprints cannot be stolen.

Using the biometric fingerprinting device requires placing a finger against a small surface or above an opening for less than 5 seconds, depending on the type of technology the scanner uses. If used for verification, it may be used along with a piece of identification.

Advantages of fingerprint scanners include the following:

- They are small and versatile. They can be integrated into numerous access points such as computer keyboards, doors, lock mechanisms, safes, and automated teller machines (ATMs).

- They are easy to use.

- The technology has existed for a long time and has high levels of accuracy. The chance of accepting an impostor is lower than the chance of rejecting a valid user.

- It is highly unlikely that two fingerprints will be alike (1 in 64 billion) even from identical twins.

- To prevent fake fingers from being used, many systems also measure blood flow or check for correctly arrayed ridges at the edges of the fingers.

Disadvantages of fingerprint scanners include the following:

- Dryness, dirt, cuts, and bruises on the finger may affect the reading.

- In sensors requiring pressing the finger to a surface, residue from a previous print may affect the reading.

- A finger mold, an image, or other replica of a finger may fool it. More devices now require a living finger. Such devices have sensors that pick up heart rate, blood flow, and temperature readings that can be evident only in a living finger.

Figure 5-4 A Fingerprint Scanner A fingerprint device scans a finger and creates a digital image.

(Benne Ochs/Getty Images)

Fingerprint scanners are currently used in:

- The City of Glendale offices in Los Angeles County, California, instead of passwords
- Universal Studios and the Minneapolis-St. Paul International Airport to access lockers
- Places needing enforced security such as doors, locks, safes and vaults, vehicles, homes, alarm controls, firearms, laptops, and cellular phones
- Biometric time-clock systems to keep track of employees
- Access sites to replace passwords for computer and network use
- Automatic teller machines (ATMs)
- Computer database and network access
- Parolee and home arrest monitoring systems
- Prisons for inmate identity verification
- Military installations for refugee and prisoner of war identification
- Enforcement agencies for identity checks
- Cable television systems for secure cable television access

b. Face Recognition Systems

A facial recognition system (Figure 5-5) may be used in a random environment to identify someone in a crowd, an example of identification. It may also be used in a controlled environment where an individual is recognized so that they may enter a site. This is an example of verification or authentication. In controlled environments, this system may be used with another piece of identification such as a card, personal identification number (PIN), password, or code.

Facial recognition systems scan a person's face by using a digital video camera. The system creates a digital image and then takes measurements of various facial structures, including the distance between the eyes, nose, mouth, and the length of the jaw, nose, eyes, and mouth. Measurements are turned into encoded data and used to compare with existing data or stored for later use.

A digital video camera is currently used only for verification. To use the system, the person faces the camera about 2 feet away while the camera locates the person's face and makes a match in less than 5 seconds. To ensure that the camera will pick up a living image, many systems require that the person make some type of facial expression such as smile or wink. However, most systems can pick up an image without the person's cooperation.

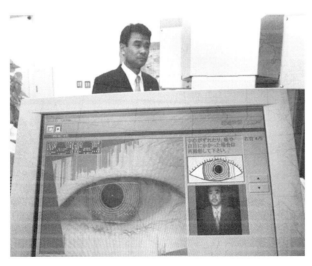

(Reuters/CORBIS)

Figure 5-5 Facial Recognition System A facial recognition system creates a digital image and takes a variety of facial measurements for verification purposes.

Advantages of facial scanners include the following:

- They are fairly inexpensive.
- They can match a face to existing records, such as a driver's license.
- The user does not have to cooperate.
- The devices are nonintrusive.
- Since people are used to being photographed, it's likely to be well accepted.

Disadvantages of facial scanners include the following:

- The type of lighting and intensity of background light may affect the image.
- Capturing varying angles of a moving face is challenging.
- They cannot distinguish identical twins.

Facial scanners are currently used in:

- Law enforcement
- Computer and network access sites
- Institutions trying to prevent terrorist crimes
- A Texas company to cash checks for customers using its automated check-cashing machines

c. Hand Geometry Systems

A hand geometry reader (Figure 5-6) is a device that measures overall hand shape, finger length, finger thickness, palm size, overall area, relationship between fingers, and other features by capturing a 3D view of the hand. It is used only for verification in combination with identification information such as a card, PIN, or password. It works by placing a hand palm-side down on a metal surface that has metal pegs to align each finger. The device then verifies the user in a matter of seconds.

Advantages of hand geometry systems include the following:

- They are easy and accurate to use.
- A small amount of data is encoded.
- They are inexpensive compared to other biometric devices.

Figure 5-6 A Hand Geometry System A hand geometry system measures hand characteristics in conjunction with identification information to provide verification.

(Jim Commentucci/Syracuse Newspapers/ The Image Works)

Disadvantages of hand geometry systems include the following:

- Hand geometry is not unique to an individual.
- They require special hardware and take up a large amount of space.
- They are difficult to integrate into other equipment.
- Current hand geometry readers have no way of detecting living tissue and may be fooled by a fake if the correct amount of pressure is applied to the plate.
- Hand geometry may vary over a lifespan.
- Arthritis, problems with dexterity, rings, and missing fingers may affect readings.

Hand geometry systems are currently used in:

- San Francisco International Airport
- University of Georgia cafeteria
- Walt Disney World
- Day care centers
- Apartment buildings
- Welfare agencies
- Hospitals
- Government facilities
- Immigration facilities
- Olympic events

d. Iris Scanners

Iris scanners (Figure 5-7) analyze the complex features of the colored part of the eye surrounding the pupil. The iris (Figure 5-8) is highly detailed, with more than 200 match points, such as rings and freckles, which can be used for comparison between individuals.

Iris scanners use a video camera to capture an image of the iris that is detailed enough to work for both identification and verification. To use it, a person must stand in front of the camera so that he or she can see his or her own reflection in the device. It takes about 5 seconds to complete the scan. The scan can be performed from a distance that can vary from 2 feet to a couple of inches, depending on the technology of the device. Most devices shine light into the eye to cause pupil dilation and to decrease the likelihood that a fake eye is used.

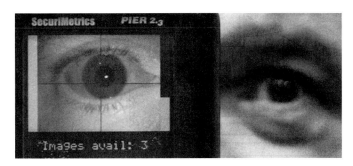

(Ian Waldie/Getty Images)

Figure 5-7 Iris Scanner An iris scanner analyzes an iris to be used for identification and verification.

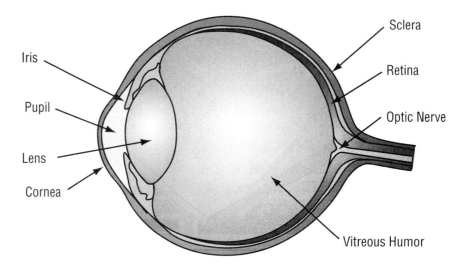

Figure 5-8 Anatomy of the Eye The iris is a very detailed portion of the eye located toward the front of the eye.

Sclera

Retina

Optic Nerve

Vitreous Humor

Iris

Pupil

Lens

Cornea

Advantages of iris scanners include the following:

- Iris patterns are unique to each individual. Left and right irises are unique within the same individual. Even identical twins have different patterns.
- Iris patterns do not vary with age.
- The user does not have to come in contact with the scanner.
- They are extremely accurate.

Disadvantages of iris scanners include the following:

- They are very expensive.
- Individuals have varying sensitivity to light.
- They require a lot of data storage.
- The accuracy decreases with glasses, fake eyelashes, lenses, or other objects that can cause a reflection.
- Specific lighting conditions are required.

Iris scanners are currently used in:

- Law enforcement agencies in the United States since 1994
- Correctional facilities
- The Charlotte/Douglas International Airport in North Carolina and the Flughafen Frankfort International Airport in Germany, both of which allow frequent passengers to register their iris scans to optimize the boarding process
- The Nationwide Building Society in Britain as a replacement for PINs at ATMs
- The National Bank United in the United States for ATM access in Houston, Dallas, and Fort Worth

e. Retinal Scanners

Retinal scanners (Figure 5-9) look at the patterns formed by the blood vessels in the retina of the eye (Figure 5-8 and Figure 5-10). The **retina** is the light-sensitive membrane covering the back wall of the eyeball and is considered the primary organ of vision because it receives images and is connected to the brain by way of the optic nerve.

A retinal scan works by shining a low-intensity light into a person's eye. It is a 360-degree circular scan of the eye, which takes more than 400 readings and

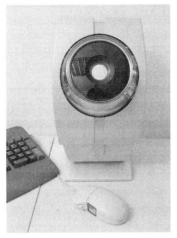

(Benne Ochs/Getty Images)

Figure 5-9 Retinal Scanner Retinal scanners analyze the patterns formed by blood vessels in the retina of the eye.

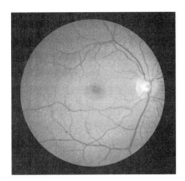

(UHB Trust/Getty Images)

Figure 5-10 Blood Vessels in the Retina of the Eye Individuals' retinas contain distinct blood vessel patterns.

stores them as encoded data used for verification. The person must remove eyeglasses if wearing any, place his or her eye close to the device, look through a small opening, and focus on a specific point.

Advantages of retinal scanners include the following:

- Retinal patterns are unique to each individual, even in identical twins.
- They are extremely accurate.
- Retinal patterns do not change with time, environmental conditions, or health of the individual.
- There is no known way to replicate a retina, and a retina from a dead person would not work.

Disadvantages of retinal scanners include the following:

- The device has to be perfectly aligned with the eye.
- They are very expensive.
- They are intrusive and can be difficult to use.

Retinal scanners are currently used in:

- High-end security facilities
- Military installations
- Power plants
- ATM machines in Japan
- Government facilities in Illinois for the prevention of welfare fraud

f. Voice Recognition Systems

Voice recognition systems (Figure 5-11) analyze distinct patterns in a person's voice to verify the identity of an individual. To use it, a person only has to speak a password or access code into a microphone hooked up to a computer system and wait for verification, which usually takes about 5 seconds. The system encodes the sound waves into a voice print and analyzes the cadence, frequency, pitch, and tone. The person's sound print is compared to a previously stored recording.

Figure 5-11 A Voice Recognition System Voice recognition systems analyze distinct patterns in speech for verification purposes.

(AJSlife/Alamy)

Voice recognition devices require a match between high and low frequencies so that a recorded voice cannot be used. Fooling the system by imitation is also difficult, because even if the impostor is very good, voice patterns are not easy to duplicate. Voice recognition is used for verification only, so it is used in combination with other identifiers.

Advantages of voice recognition systems include the following:

- They are easy to use.
- They are easy to install because a microphone can be easily connected to a computer.

Disadvantages of voice recognition systems include the following:

- Background noise interferes with the analyses.
- They have a high error rate.
- Software is expensive.
- Extensive training is required prior to using the software.

Voice recognition systems are currently used in:

- Computer and network sites
- Financial institutions

g. Signature Recognition Systems

Biometric signature systems (Figure 5-12) analyze the features of a signature as well as the physical activity during the signature. For example, it measures the pressure on the writing surface, the time it takes to write each letter, the total signature time, whether *t*'s are crossed and *i*'s are dotted, and many other characteristics. It is easy to use, since people have been signing their names for centuries for the purposes of authentication. The signature has been a common form of authentication for many years.

Signature systems have the ability to distinguish between a person's usual patterns of signing and the variations that take place every time a signature is made. Usually, a person will sign on a tablet or on a special sheet of paper over a sensor. The device analyzes the signature and verifies it in about 5 seconds. This system is almost always used with other identifiers.

Figure 5-12 Signature Recognition Tablet Signature systems analyze the features of a signature along with physical activity during the signing for verification purposes.

(Ronald Karpilo/Alamy)

Advantages of signature recognition systems include the following:

- They are easy to use.
- A signature is difficult to forge.
- They integrate well with other systems.
- They are noninvasive.

Disadvantages of signature recognition systems include the following:

- They have limited applications.
- Signatures change over time.
- People who are inconsistent with their signatures lead to high error rates.
- Professional forgers may be able to fake a signature.

Signature recognition systems are currently used in:

- Financial transactions

h. Keystroke Dynamics

Also called **typing rhythms**, this method looks at the way a person types by analyzing keystroke patterns and speed. It makes approximately 1000 analyses per second. Both the National Science Foundation and the National Institute of Standards and Technology have conducted studies that have concluded that keystroke dynamics are unique to each individual. This biometric system does not require a special scanning sensor. The computer does the analysis directly, so it requires only special software.

This system is usually used for verification as a user types a password or other code. Verification takes less than 5 seconds. Research into keystroke dynamics is ongoing.

i. Future Biometrics

While the biometric tools discussed previously are currently in use, other biometric tools are being developed and researched. Some of those include:

- Vein pattern reader: Analyzes the blood vessel pattern on the back of the hand. Vein patterns are thought to be unique even between identical twins. They also persist throughout the lifetime of an individual.
- Facial thermography device: Captures the patterns of facial heat caused by blood flow below the skin by using an infrared camera. Facial thermographs are considered unique but not stable under varying environmental conditions.
- Sweat pores analysis: Analyzes sweat pore distribution patterns to identify a person. Sweat pores are responsible for **latent fingerprints**

left behind by an individual (see Project 2 for more on latent fingerprints). This method is experimental.

- Hand grip device: Measures patterns made by a person's grip by using two different methods. In one method, an infrared light illuminates the skin and looks at the underlying blood vessel patterns made upon gripping. The other method analyzes the dynamics of the bones, muscles, and joints while gripping an object.

- Fingernail bed analysis: Captures the patterns in the skin underneath the fingernails. This skin is made up of unique patterns of parallel lines formed by specific skin structures. It is like a personal barcode.

- Body odor device: Captures the unique scent of an individual, composed of about 30 chemicals that form a unique pattern. The device would be like an electric nose, with special receptors for these scents. This technique is young in its development and may not be very reliable due to environmental causes for changes in odor.

- Ear shape analysis: Measures the ear's overall shape and geometry in additional to in-depth analyses of the ridges of ear tissue. The ear has been considered a good measure of identity since the time of Bertillonage.

- Gait analysis: Measures and models pendulum motion of the legs. It would also consider motion of the whole body. Determined by weight and height, gait is the distinctive way a person walks. Research has shown that gait movements are unique to an individual.

- Skin luminescence analysis: Analyzes the patterns of reflected light as a light is shone through the skin. This creates a unique pattern that is thought to be stable.

- Brainwave pattern device: Measures and records baseline brainwave patterns created by the electrical signals in the brain (electrical potentials) caused during brain activity. Unlike most brainwaves, baseline brain waves cannot be altered.

- Footprint and foot dynamics: Analyzes the footprint, similar to a fingerprint, for ridges and other structures. It would also analyze the dynamics of a footstep as a person walks by and evaluate step size, pressure, timing, and friction. This method is only hypothetical at this point.

- Deoxyribonucleic acid (DNA) biometric device: Automates DNA analysis within a single unit for immediate analysis of a person's DNA.

Key Terms

biometric	fingerprint scanners	signature recognition systems
fingerprints	face recognition systems	keystroke dynamics readers
identification	hand geometry readers	minutiae
verification	iris scanners	retina
authentication	retinal scanners	typing rhythms
biometrics device	voice recognition systems	latent fingerprints
match points		

III. Student Materials

Use Project 1 Design Brief: Biometric Tools to complete the project.

UNIT 5: PROJECT 1   DESIGN BRIEF: BIOMETRIC TOOLS

Your class has been asked to develop a Web site about biometric tools. Your job is to become an expert on one particular biometric device by researching how it works, how and where it is used, and its advantages and disadvantages. You also need to design a future application for your biometric device.

Design Constraints: Things You Must Do

Below are constraints, or guidelines, to follow during this design brief:

- Define biometrics, give historical background, and give information about your biometric tool.

- Collect information about your biometric tool's current applications in the United States as well as in other countries.

- List the major manufacturer(s) of your biometric tool.

- Discuss the advantages and disadvantages associated with your biometric tool and compare it to other biometric devices.

- Create a graphic that describes the underlying technology and functionality of your biometric tool.

- Design a new application for your biometric tool.

- Discuss the following issues with respect to the new application:

    - Performance: How well does it perform in its new use?

    - Universality: Can people with physical/learning disabilities use it?

    - Security: How easily will it accept an impostor?

    - Validity: How often will it reject the authorized user?

    - User acceptance: To what extent are people willing to accept it?

PROJECT 2
Collecting and Analyzing Fingerprints

I. Project Lesson Plan

1. Project Description

You will explore the most common biometric measure—the fingerprint. You will learn how to collect, analyze, and match fingerprints using existing classification systems.

2. Learning Objectives

- You will study the anatomy of a fingerprint by analyzing various fingerprints.
- You will learn how to collect your own fingerprints.
- By comparing your own fingerprints to others, you will learn how to classify fingerprints and make a match between two fingerprints.
- Using fingerprint patterns and minutiae points as your data sources, you will make graphs of fingerprint pattern classroom distributions and your own minutiae point frequencies.

II. Background Information

1. Fingerprint Biometrics

Fingerprints are the oldest and most common **biometric** used to identify individuals. Fingerprints are impressions, or prints, made from **ridges** (i.e., lines going across) and **valleys** (i.e., spaces between ridges) on our finger pads. The pattern of ridges and valleys is unique—no two individuals, even identical twins, have fingerprints that are exactly alike. The chance of two people having the same print is less than 1 in 1 billion. Even injuries such as burns or scrapes will not change the ridge structure because when new skin grows in, the same pattern will return. The uniqueness of fingerprints makes them an excellent biometric for identifying individuals.

We leave fingerprints everywhere, on everything we touch. If your fingers are dirty or oily, you can see your fingerprints when you touch something. These are called **visible prints**. If your hands are clean, prints are made by sweat that is naturally produced by sweat glands present on the fingertips. These prints are known as **latent prints** and can be made visible by various techniques. Another type of print is called an **impression**. These fingerprints are made by contact with a soft, pliable surface such as clay, wax, wet paint, blood, or anything else that may take an impression.

2. Taking and Lifting Fingerprints

The Federal Bureau of Investigation (FBI) and other law enforcement and government agencies usually take fingerprints by pressing fingers to an inkpad and then transferring the inked fingers to a fingerprint card (Figure 5-13). The most common type of fingerprint card has spaces for individual fingerprints as well as spaces for simultaneous prints of all fingers. The fingerprints are taken by rolling each individual finger from edge to edge and by pressing (or "slapping") simultaneous fingerprints. The thumb is printed twice—once by rolling and once by slapping. Fingerprint cards are catalogued and stored for later use.

Not all prints are taken voluntarily. When left at crime scenes, fingerprints may be either visible or latent. When fingerprints are latent, various methods are used to detect or "develop" the print, such as chemicals, powders, lasers, light sources, and other techniques (Figure 5-14). The method used depends on how the print was formed.

Once the fingerprint is developed, it is usually photographed and lifted. Lifting techniques are used for any crime scene print that needs to be removed for analysis and storage. See Figure 5-15 for an example of a typical fingerprint lifting

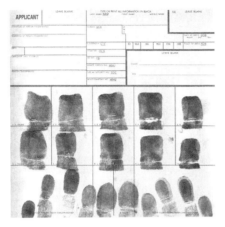

(Randy Faris/CORBIS)

Figure 5-13 A Typical Fingerprint Card Most fingerprint cards have spaces for individual fingerprints and simultaneous prints of all fingers.

(Stephen Vowles/Alamy)

Figure 5-14 Dusting with Powder to Detect a Latent Print Forensics units have several techniques to develop a fingerprint, including dusting with powder.

(Linda McPherson/Fotolia)

Figure 5-15 Fingerprint Kit Fingerprint lifting kits remove fingerprints left at crime scenes for analysis and storage.

kit. The type of surface where the fingerprint was found determines the lifting technique. Although there are various lifting devices, high quality cellophane tape is used often, especially when lifting fingerprints from flat surfaces.

3. Matching Fingerprints

Fingerprints are unique because they each have a distinct pattern created by the ridges on the skin and by the valleys those ridges form. The types of ridge patterns, the size of those patterns, and the position of the patterns on the finger classify fingerprints. Ridges and valleys tend to be uneven, with breaks and interruptions across our fingers, hands, toes, and feet. This uneven sequence of ridges and valleys is characterized by **minutiae** (plural) or **minutia** (singular). Minutiae are composed of a variety of sweat pores, distance between ridges, bifurcations, junctions, and endpoints. There are about 100 minutiae points in an average finger, but usually only about 30 to 60 are captured due to the size of the print and the area of the device used to capture the print.

Several categories of minutiae are listed in Figure 5-16 and Table 5-1, but usually ridge endings and bifurcations tend to be the focus of fingerprint analyses

Figure 5-16 Representation of Minutiae Minutiae are composed of a variety of sweat pores, distance between ridges, bifurcations, junctions, and endpoints.

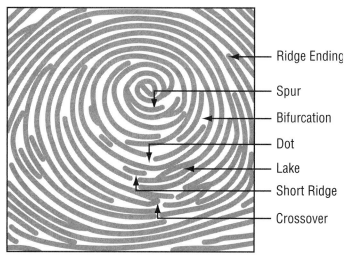

(Courtesy of The CrimeTrac Agency)

Table 5-1 Minutiae in More Detail

Crossover (bridge)		Double bifurcation	
Trifurcation		Opposed bifurcations	
Ridge crossing		Opposed bifurcation/ ridge ending	
Dot		Ridge ending	
Short ridge (island)		Bifurcation	
Lake (enclosure)		Hook	

Table 5-2 Fingerprint Patterns Used in Classification

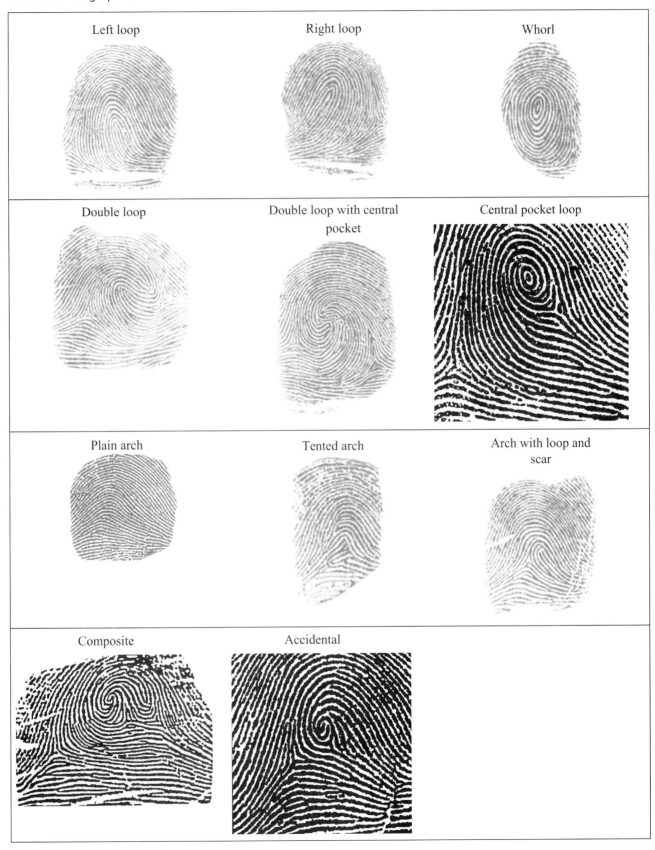

Left loop   Right loop   Whorl

Double loop   Double loop with central pocket   Central pocket loop

Plain arch   Tented arch   Arch with loop and scar

Composite   Accidental

Note: This system of classification is based on the Henry System, named after Sir Edward Henry, who established a systematic approach for fingerprint classification (see Project 1 for more historical information).

since other minutiae are composed of these two main types. Minutiae used for comparisons of fingerprints are referred to as minutiae points.

The types of minutiae used in analysis include:

- Ridge ending: a ridge that ends abruptly
- Bifurcation: a single ridge that divides into two ridges
- Lake or enclosure: a single ridge that bifurcates and reunites shortly afterward to continue as a single ridge
- Short ridge, island, or independent ridge: a ridge that commences, travels a short distance, and then ends
- Dot: an independent ridge with approximately equal length and width
- Spur: a bifurcation with a short ridge branching off a longer ridge
- Crossover or bridge: a short ridge that runs between two parallel ridges

It is important to emphasize that no two prints are identical, even from the same person. For example, how much you press or roll your finger on a surface can change from print to print. Also, how sweaty, dirty, or chapped your fingertips are will affect the print.

When matching fingerprints, we rely on two main categories:

1. Classification: to classify prints into broad categories (e.g., the Henry System discussed in Project 1) based on overall appearance
2. Analysis: to focus and map minutiae patterns

Minutiae patterns usually fall into basic classes: right loop, left loop, whorl, arch, and tented arch. Combinations of these patterns are classified into additional categories (see Table 5-2 for more information). Fingerprints are usually compared to each other on the basis of minutiae type, number, direction, and position. If a sufficient number of minutiae are located in two prints, they are considered points of similarity, leading to a match. Currently, there are no international standards for the number of points of similarity between two fingerprints (Project 3 discusses fingerprinting devices that analyze and compare fingerprints).

Key Terms

fingerprints	valleys	impression
biometric	visible prints	minutiae
ridges	latent prints	minutia

III. Student Materials

Use Project 2 Design Brief 1: Collecting and Analyzing Fingerprints and Project 2 Design Brief 2: Lifting Latent Fingerprints to complete the projects.

UNIT 5: PROJECT 2   DESIGN BRIEF 1: COLLECTING AND ANALYZING FINGERPRINTS

You have been promoted to the forensics team at your city's police department. As part of your training, you need to learn how to collect, analyze, and match fingerprints. Conduct the following training activity carefully. You will have to turn in your analyses, including a graph.

Design Constraints: Things You Must Do

Below are constraints, or guidelines, to follow during this design brief:

1. Work with a partner for this activity.

2. Rub a pencil over the central part of an index card until you have an area the size of a quarter. This will be your ink pad.

3. Press your index finger onto the ink pad and roll from side to side gently. Make sure that you press the part closest to the joint crease and not the very tip of the finger.

4. Take some tape and gently press it onto your index finger to lift the print from your finger. The print should now be on the tape.

5. On the tape with the print on it place another index card and label it with your name, the finger (e.g., index finger), and whether it is your left or right finger.

6. Repeat Steps 3–5 with all other fingers until you have a full set of fingerprints.

7. After all prints are made and labeled, classify your prints into one of the fingerprint pattern categories. (See the classification chart in Table 5-2 on page 192.)

   When matching fingerprints, we rely on two main categories:

   • Classification: to classify prints into broad categories based on overall appearance

   • Analysis: to focus and map minutiae patterns

8. Once you have classified your prints according to general pattern, look for minutiae. Use the information in Figure 5-16 and Table 5-1 on page 191 to help you analyze and identify different types of minutiae. Label them on your fingerprints, and keep a count of each type.

9. Graph the number of different types of minutiae you found on your fingerprints. Your graph should show frequency of minutiae.

UNIT 5: PROJECT 2   DESIGN BRIEF 2: LIFTING LATENT FINGERPRINTS

The second phase in your forensics training is to learn how to collect latent prints. You arrive at a "crime scene"—your kitchen or bathroom (some place easy to clean up!)—and you have to detect and develop latent fingerprints at the site. Once you are done, you must classify them. Create a small presentation showing how you found, developed, and matched the print.

Design Constraints: Things You Must Do

Below are constraints, or guidelines, to follow during this design brief:

1. Gather all your materials:

   • Fingerprint powder

   • Paintbrush

   • Tape

   • Shiny black paper

2. Choose a "suspect." It can be yourself or a friend or family member. Ask them to touch things at the "crime scene" while you wait outside.

3. Take your "suspect's" fingerprints using the technique you learned in Design Brief 1.

4. Now, go into the "crime scene" and dust for fingerprints. Make sure to dust lightly so that you do not disturb the crime scene. Blow excess powder gently. Very carefully, brush the powdered spots with the paintbrush until the fingerprint shows. This may take some practice.

5. To keep the fingerprint, press a piece of tape over it and peel it away with the powdered fingerprint on it. Stick the tape on the shiny black piece of paper so that it can be seen more easily.

6. Collect as many different prints as you can.

7. Compare the prints you collected to the "suspect's" fingerprint. See if you can make a match.

8. Put together a formal presentation with images detailing your techniques and procedures.

PROJECT 3
Processing Fingerprints

I. Project Lesson Plan

1. Project Description

This project introduces you to fingerprinting applications and processing techniques associated with various fingerprint biometric devices. As part of the design brief, you will match a crime scene fingerprint with a suspect's fingerprint by using image processing and highlighting techniques.

2. Learning Objectives

- You will learn about various fingerprint biometric devices, how they work, how they are used, and their strengths and weaknesses.
- You will learn how to enhance and compare fingerprint images by using image processing software.
- You will explore various forms of conceptual data representation by presenting fingerprint collection methods and image processing results.
- You will assess the advantages and disadvantages of various fingerprint biometric devices with respect to image quality.

II. Background Information

1. Fingerprint Biometric Devices

Biometric devices that use **fingerprints** as the measure for identification have a 3 percent probability that the system will fail by falsely rejecting an individual. However, the probability of a false acceptance is less than 1 in 1 million. A typical automatic **fingerprint scanner** requires the user to place a finger on the machine for as little as 0.5 second to 2 seconds. Figure 5-17 shows an example of a fingerprint access control panel.

The following are several technologies used in fingerprint devices.

- Capacitive method: This is one of the most popular technologies. A scanner with specialized capacitive sensors generates an image of a fingerprint's **ridges** and **valleys**. The capacitive sensor uses capacitors

(Jon Feingersh/zefa/Corbis)

Figure 5-17 Example of a Fingerprint Access Control Panel The fingerprint access control panel requires the user to place a finger on the machine for 0.5 to 2 seconds.

(i.e., electrical devices that can store electric charges) to measure the fingerprint. Although this system is effective, it requires a live fingerprint as opposed to an image. Consequently, it may be difficult for the sensors to pick up prints from fingers that are either too moist or too dry, which result in smudgy or pale images, respectively.

- Thermal-electric method: This method is far less common than others. It uses a sweeping action across a sensor, which measures temperature differences between a finger's ridges and valleys. Fingerprints are captured in a series of image slices. Then, using special software, the slices can be put together to create a high quality image (500 dots per inch with 256 grayscales), even from poor quality fingerprints. Unlike the capacitive method, it is not sensitive to variations in temperature, humidity, or skin moisture. The disadvantage of this method is that it requires some skill in using the scanner.

- E-Field sensor: This device measures the electric field beyond the surface layer of the skin where the fingerprint begins. It is a hardy system, working in real-world conditions and on dry, worn, or dirty skin.

- Optical method: This device uses an optical scanner to scan a finger placed on a glass plate where a specialized camera takes a picture. This technology is fairly inexpensive, but it is easy to fake because the scanner may confuse a picture of a finger with a real finger. Another disadvantage is that it tends to be contaminated by fingerprints that might remain on the sensor from a previous user.

- Surface-pressure sensor: This sensor works by picking up the finger ridges that come in contact with a sensor, leaving out the valleys. It offers high accuracy of recognition, leading to a low error rate while working with both dry and wet fingers.

- Touchless sensor: This is similar to the optical method. It works by scanning a finger, which is placed over an area with an opening that has a sensor beneath it. One disadvantage is that debris can fall into the opening and reduce the quality of the images.

Despite the advances in the technology of fingerprint biometric devices, there continue to be some problems associated with capturing clear, well-defined fingerprints. They include:

1. Positional: The way a person positions his/her finger on or over the device may cause a rotation or an incorrect translation of the print when the print is scanned.

2. Pressure difference: The amount of pressure an individual places on a device may reduce uniformity in the print, creating areas that are difficult to analyze. Touchless sensors, as discussed earlier, may minimize this problem.

3. Obstruction: Oily, soiled, dry, scarred skin may cause an obstruction, as can dirt or residue on a device's sensor.

There are two kinds of fingerprint biometric systems: **Automatic Fingerprint Identification Systems (AFIS)** and **Automatic Fingerprint Authentication Systems (AFAS)**. AFIS have historically been used by law enforcement agencies, commonly in crime scene investigations (CSI). In this system, the input is the fingerprint, and the output is a match with the person to whom the fingerprint belongs. It compares the input fingerprint with a database of fingerprints to make the match.

In AFAS, the input is an ID, for example, an ID card, plus a fingerprint. The output is either a "yes, you are who you say you are," or "no, you're not who you say you are." It works by matching the fingerprint to the ID provided. If the two pieces of information match, then the user is verified. In this manner, personal identification numbers (PINs), passwords, or multiple ID cards, which may be easily lost or stolen, are matched to a unique biometric such as a fingerprint to safeguard their appropriate use.

An example of an AFAS is a smart card, which is a card that looks like a credit card but has an embedded 8-bit microprocessor, coupled with live fingerprint scanning. In this case, a digital template of the user's fingerprint may be captured and stored with identifying information on a smart card. When an individual inserts the card into a given terminal at a bank, door, or computer system, he or she is prompted to place a finger on a fingerprint scanner to determine if the fingerprint matches the identifying information stored on the card.

These systems are currently expanding to identify those who claim welfare benefits; to monitor prisoner movement; to secure credit cards or bank cards; to replace or complement password usage in computer systems; to secure personal computers, PDAs, and cell phones; and to confirm memberships and passport holders at border crossings and at military and research facilities. Figures 5-18 through 5-21 show some fingerprint biometric devices.

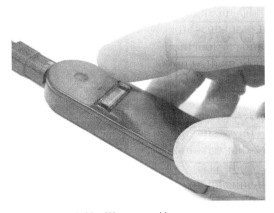

(Gary James Calder/Shutterstock)

Figure 5-18 USB-style Fingerprint Reader This is a computer mouse fingerprint reader with an optical system for computer and network users seeking security.

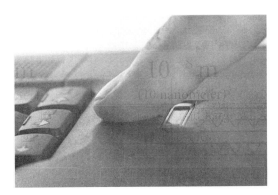

(© pm/Alamy)

Figure 5-19 Fingerprint Scanner Keyboard The keyboard uses fingerprint minutiae recognition to authenticate users who want access to a network server, personal workstation, or personal computer.

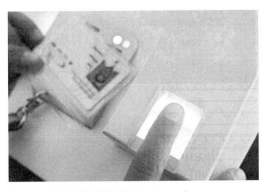

(JOHANNES EISELE/AFP/Getty Images)

Figure 5-20 Smart Card System This is an AFAS system using smart card technology for access control.

Figure 5-21 Personnel
Identification This device
features a biometric one-touch
system that can store up to
4000 finger-scan templates for
verification and optional
dual-factor authentication.

(Richard Wadey/Alamy)

2. Fingerprint Image Processing

Comparison of fingerprints relies on analyzing the **minutiae**, which are the ridges
on the skin and the valleys those ridges form (see Project 2 for more information on fingerprint analysis). Comparison and analysis of a fingerprint require a
good quality print, but prints may be contaminated or incomplete. Image processing techniques are used to enhance low quality fingerprints taken by fingerprint
biometric devices or lifted from crime scenes. By using color and contrast, for
example, image processing techniques clarify, fill in gaps, bring out details, and
highlight specific points.

The first step in enhancing a fingerprint is to make it stand out by adjusting contrast and brightness. Working with contrast and brightness may allow a print to pop
out from its background (Figure 5-22) so that you can work with it more easily.

Once the image is visible, you can use other image processing tools to classify the fingerprint. First, you can look for overall patterns to help you classify the
fingerprint and use various tools such as lines or color to help you highlight the
patterns (Figure 5-23).

You can highlight more complex overall patterns by using different lines or
colors. For example, in Figure 5-24, lines are used to show the two loops in a
double-loop pattern. You could also use a different color to show each loop.

To highlight minutiae points, you can use color, labels, tags, boxes, or lines
when comparing prints (Figure 5-25).

Figure 5-22 Fingerprint
Enhancement through
Contrast Adjustment The
image on the left is prior to
adjusting the contrast. The image
on the right shows the fingerprint
after contrast adjustment.

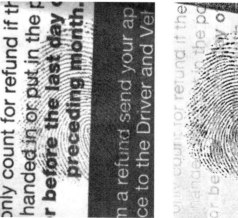

(Both images courtesy of Neate Imaging Services)

Figure 5-23 Using Lines to Highlight a Fingerprint Pattern By outlining the dominant pattern, the loop in this fingerprint becomes easier to detect. The pattern could also be highlighted in a contrasting color.

(Courtesy of Neate Imaging Services)

Figure 5-24 Using Two Lines or Colors to Highlight a Pattern Using two lines or colors makes it easier to distinguish complex patterns.

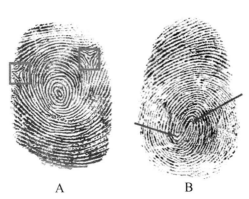

A   B

Figure 5-25 Examples of Highlighting Techniques for Comparing Minutiae Print A shows the use of arrows to point at specific features. Print B shows minutiae identification by boxing.

3. Fingerprint Characteristics

Since human skin fluctuates depending on environmental factors, no two prints, even from the same person, are ever identical. Therefore, the biometric devices that analyze fingerprints work through a series of **algorithms**—step-by-step procedures for solving a problem—that look for patterns in minutiae.

Minutiae patterns usually fall into basic classes:

- Right loop
- Left loop
- Whorl
- Arch
- Tented arch

Combinations of these patterns are classified into additional categories. Fingerprints are usually compared to each other on the basis of minutiae type, number, direction, and position.

A biometric fingerprint device will record minutiae points on a coordinate system and either create a template of the fingerprint or compare it to an existing template. The template is like a map, containing all the minutiae information. It may have more information than just the fingerprint to increase the probability of an accurate comparison. Most biometric devices store that map only in their databases so that the actual image of a print cannot be copied for criminal uses. If a sufficient number of minutiae are located in two prints, they are considered points of similarity, leading to a match. Currently, there are no international standards for the number of points of similarity necessary to declare a match between two fingerprints.

The types of minutiae used in analysis include:

- Ridge ending: a ridge that ends abruptly
- Bifurcation: a single ridge that divides into two ridges
- Lake or enclosure: a single ridge that bifurcates and reunites shortly afterwards to continue as a single ridge
- Short ridge, island, or independent ridge: a ridge that commences, travels a short distance, and then ends
- Dot: an independent ridge with approximately equal length and width
- Spur: a bifurcation with a short ridge branching off a longer ridge
- Crossover or bridge: a short ridge that runs between two parallel ridges

Figure 5-26 Representation of Minutiae Minutiae are composed of a variety of sweat pores, distance between ridges, bifurcations, junctions, and endpoints.

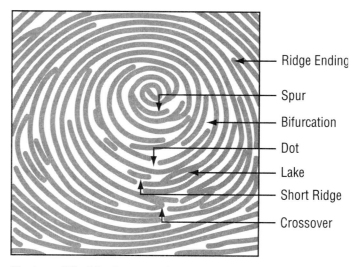

Ridge Ending

Spur

Bifurcation

Dot

Lake

Short Ridge

Crossover

(Courtesy of The CrimeTrac Agency)

Table 5-3 Minutiae in More Detail

Crossover (bridge)		Double bifurcation	
Trifurcation		Opposed bifurcations	
Ridge crossing		Opposed bifurcation/ ridge ending	
Dot		Ridge ending	
Short ridge (island)		Bifurcation	
Lake (enclosure)		Hook	

Table 5-4 Fingerprint Patterns Used in Classification

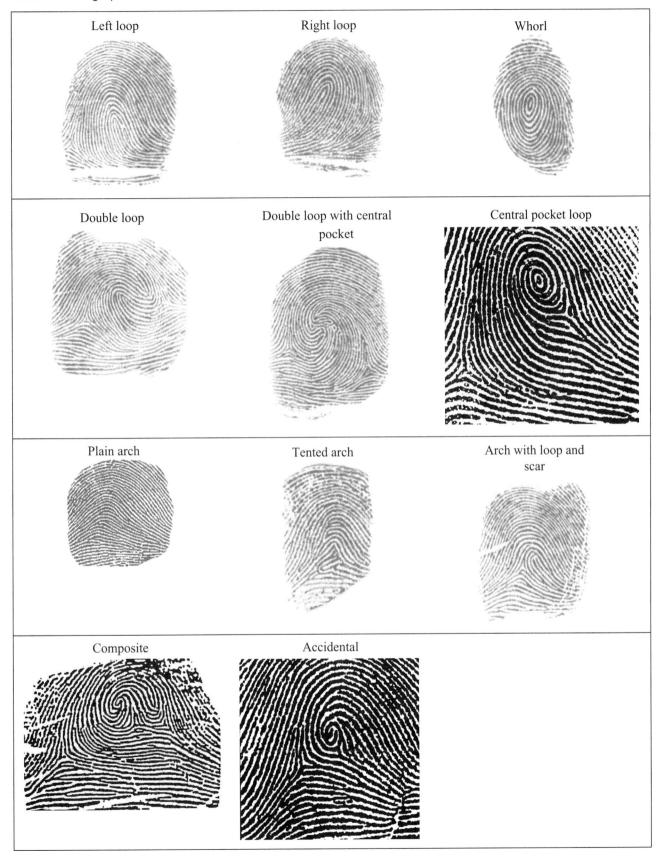

Left loop	Right loop	Whorl
Double loop	Double loop with central pocket	Central pocket loop
Plain arch	Tented arch	Arch with loop and scar
Composite	Accidental	

Note: This system of classification is based on the Henry System, named after Sir Edward Henry, who established a systematic approach for fingerprint classification (see Project 1 for more historical information).

biometric device
fingerprints
fingerprint scanner
ridges
valleys

Automatic Fingerprint
   Identification Systems (AFIS)
Automatic Fingerprint
   Authentication Systems
   (AFAS)

minutiae
algorithm

III. Student Materials

Use Project 3 Design Brief: Fingerprint Characteristics to complete the project.

UNIT 5: PROJECT 3   DESIGN BRIEF: FINGERPRINT CHARACTERISTICS

Identify visually the fingerprint characteristics of the prints below. You may want to use a magnifying glass.

The types of minutiae used in analysis include:

- Ridge ending: a ridge that ends abruptly

- Bifurcation: a single ridge that divides into two ridges

- Lake or enclosure: a single ridge that bifurcates and reunites shortly afterwards to continue as a single ridge

- Short ridge, island, or independent ridge: a ridge that commences, travels a short distance, and then ends

- Dot: an independent ridge with approximately equal length and width

- Spur: a bifurcation with a short ridge branching off a longer ridge

- Crossover or bridge: a short ridge that runs between two parallel ridges

You can refer to Figure 5-26 on p. 201 and Tables 5-3 and 5-4 on pages 201 and 202 to help.

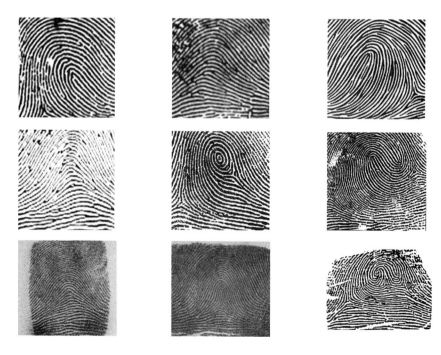

PROJECT 4
Identity Theft and Privacy

I. Project Lesson Plan

1. Project Description

You will discuss the implications of using biometric devices in our society, including issues of identity theft and privacy. In the design brief, you will take a stance for or against the use of biometrics as it relates to identity theft and personal privacy.

2. Learning Objectives

- You will develop ideas about how specific biometric devices will affect society, including legal and political issues.
- You will address the advantages and disadvantages of biometrics with respect to identity theft and privacy.
- You will address issues of identity theft and privacy within your own school system and community.
- You will support your stance on biometrics by creating multimedia presentations.

II. Background Information

1. Identity Theft

To date, all **authentication** systems can be defeated in some way. For example, passwords can be stolen and reused. To minimize the chances of being intercepted, pieces of identification may come in pairs: an ATM card and a PIN, or an ID card and a biometric. This pairing makes it more difficult for someone who steals the information to actually use it.

Despite these safeguards, the system can still be fooled in various ways. In fact, most authentication systems now in place use limited pieces of information, for example, a single identifier such as a driver's license. Faking a simple piece of identity and engaging in **identity theft** is not too difficult. An identity thief will try to trick the system into believing that he or she is the victim. If successful, the thief will have access to all of the victim's confidential files, information, and financial records. The thief can exploit those materials to hide behind the victim's identity for the purposes of committing a crime or rob the victim himself or herself. In the latter case, the thief sets up accounts under the victim's name by using authentication information stolen from the victim, such as name, social security number, mother's maiden name, or birth date.

One of the most common ways a thief can trick the system is by using a trial-and-error attack. This scenario is common with passwords. In this case, a thief will type one possible password after another, keeping track of all the combinations that have been tried. He or she can even use the help of a computer to hack an encrypted password on the Internet. In this approach, the computer does a search of words likely to be used as passwords. A thief may also try replication, where he or she will produce a copy of whatever the victim uses for identification. Then, the victim's password is stolen from a wallet or other personal item where the password may be written.

There is no obvious way for someone to use trial-and-error to recreate a biometric, but there may be ways to fool a biometric sensor by replication. The type of replication depends on the sensor and the type of verification protocol it uses.

Attacks involving replication of a biometric can be classified as **imitations** or **artifacts**. In an imitation attack, the thief impersonates another person. In an artifact attack, the thief presents a replication of the **biometric** such as a mold of a finger with a fingerprint impression from the victim. While this type of theft is possible, it is difficult to physically steal a biometric. More and more **biometric devices** are integrating sensors sensitive to live functions such as blood flow, heart rate, and body temperature. This is often referred to as the *liveness factor* of a biometric device (see Project 3 for more information on biometric devices).

Although common in movies, it is unlikely that someone will cut off a finger to use to steal someone's identity via a biometric device. More commonly, a thief may steal biometric information stored on a database. When biometric information is recorded, it is usually encrypted for storage. A thief may intercept the system as the victim is providing his or her biometric information, or the thief may hack into the system and break the encryption. This is a sophisticated type of attack and requires a certain degree of expertise and knowledge about the database system used to store the biometric information.

2. Privacy

The issue of **privacy** has come into the forefront in recent years. For example, in 2002, the Office of Homeland Security announced its "Specifics of Secure and Smart Border Action Plan," which lists biometrics at the top of the list. Following this announcement, President George W. Bush referred to biometrics in his State of the Union Address, when he said, "We will. . . use technology to track the arrivals and departures of visitors to the United States." Since then, airports around the world have begun to track foreign travelers using biometric devices to supplement legal documents such as passports. While biometrics may be used to protect a nation's boundaries, it may do so at the expense of individual personal boundaries.

When people provide identifying information about themselves, they are often concerned about who is using it, why, how, and when. Moreover, there is a difference between voluntarily giving up a biometric and having to surrender it to comply with federal law. For example, a person may willingly give biometric information to obtain a frequent visitor's pass to Walt Disney World, but may not want to be forced to provide biometric information at an airport.

These issues deal with personal freedoms. To whom do we give control over information that is literally a part of us? While we can make up a name, an address, a phone number, we cannot make up a biometric about ourselves. Does this mean that we lose the right to choose anonymity? Right now, there are no regulations in place for the storage or treatment of biometric information. Until regulatory action is in place, the use of biometrics is left to the private sector.

Although biometrics appears more intrusive than traditional forms of identification, the benefits of using it may outweigh the negatives. For example, ATM fraud, which is on the rise, leads to about $500 million in losses per year. Check fraud is also on the rise at about $2 billion in losses per year. Credit card fraud is estimated at $1.5 billion per year. Financial industries believe that these losses can be significantly reduced by the use of biometrics. Biometrics can ensure that only the authorized account holder can access the account.

Those in favor of biometrics argue that individual privacy will not be more compromised by biometrics than it already is. Currently, private and public sectors have access to substantial amounts of information about individuals. For example, credit card use, demographic factors, and health records are all available to those who want to gather personal information about an individual. Nevertheless, there are many people who prefer to make transactions and go about their daily business without leaving behind recordable pieces of identification.

Those against biometrics argue that over time, biometric information would be used for purposes not originally intended. This has already happened with other identifying information. For example, in 1941, the United States used confidential records to identify all foreign-born and American-born Japanese living in the United States. As a result, military personnel detained and relocated persons of Japanese descent, including U.S. citizens, to special camps. Considering this example, how might biometric data be used in the future?

Key Terms

authentication	artifacts	biometric devices
identity theft	biometric	privacy
initiations		

III. Student Materials

Use Project 4 Design Brief: Identity Theft and Privacy to complete the project.

UNIT 5: PROJECT 4   DESIGN BRIEF: IDENTITY THEFT AND PRIVACY

Companies involved in the development of biometrics technology are interested in how biometrics will affect our society in the future. As part of your job for the legal department of a biometric company, you have been assigned to investigate the issues of identity theft and privacy with regards to biometrics technology.

Design Constraints: Things You Must Do

Below are constraints, or guidelines, to follow during this design brief:

- Give an introduction to current issues of identity theft/privacy.
- Discuss how different biometric devices may or may not prevent identity theft or how intrusive/non-intrusive they are on individual privacy.
- Discuss how biometrics may be used in the future and how this may affect our society with respect to identity theft/privacy.
- Include all the previous points in a multimedia presentation.

PROJECT 5
Crime Scene
Visualization

I. Project Lesson Plan

1. Project Description

You will use 3D and 2D visualization software tools to re-create a crime scene. This scene will be used to analyze the current knowledge and evidence pertaining to the crime, including positions of individuals over time, line of sight, and proximity to evidence. These visualizations can be crucial for locating biometric-based evidence collected at the crime scene in space and time and re-creating its role in the crime.

2. Learning Objectives

- You will learn about the application of 2D and 3D visualization software in crime scene re-creation.
- You will explore modeling and visualization tools and techniques needed to re-create a crime scene.
- You will learn which elements of a visualization are most useful in problem solving about the crime scene.

II. Student Materials

Use Project 5 Design Brief: Crime Scene Visualization to complete the project.

UNIT 5: PROJECT 5   DESIGN BRIEF: CRIME SCENE VISUALIZATION

Your class has been asked to help solve a high-profile crime through the use of crime scene visualization techniques. To do this, you must:

- Choose a crime scene. You may research an actual crime from the newspaper or TV news, or you may borrow one from a television crime drama or film.
- Research the facts about the crime scene, including evidence type and location. This includes how events unfolded over the course of the crime.
- Create sketches of the crime scene, including evidence location.
- Create 2D and 3D visualizations of the crime scene, including hypotheses of how the crime unfolded over time. You may sketch by hand or use 2- and 3-D modeling software.

- Label on the crime scene visualizations where evidence was found, what kind of evidence was found, and how the evidence was collected.
- Create a presentation, explaining the crime:
  - What was found at the scene
  - Who was there
  - What biometric traits were found and where
  - How it happened

PROJECT 6
Biometric Device Design

I. Project Lesson Plan

1. Project Description

You will take what you have previously learned about biometric concepts and techniques and apply it to the design of a biometric device. This device will both represent the science of biometrics and meet a known need for biometric measurement and analysis.

2. Learning Objectives

- You will gain in-depth knowledge about one area of biometrics
- You will apply engineering design principles to merge biometric technology with end-user needs.

II. Student Materials

Use Project 6 Design Brief: Biometric Device Design to complete the project.

UNIT 5: PROJECT 6   DESIGN BRIEF: BIOMETRIC DEVICE DESIGN

Your company, ACME Bio-Systems, has decided to enter the biometric systems market. Your job is to research the current areas of biometrics and the commercial devices that are already on the market. From this research, you will propose a device to bring to market for your company.

For your presentation to the vice president for marketing, you will need to:

- Choose an area of biometrics to research. This should include background on the concepts and technology in this area of biometrics and existing products and applications.

- From this research, decide on a product that would either fit a niche that is not well-represented with existing products or an area of high need. Write up a proposal that includes this background research and a needs statement that demonstrates what is unique about your product.

- Create ideation sketches of at least three possible designs and have them reviewed by your teacher.

- Create either physical or virtual (computer) models of one of your proposed designs. Include supporting text and graphics explaining its functionality.

UNIT 1
Communications Technology: Introduction to Visualization

Project 1: Interpreting Graphics

Web Sites

- Neolithic Art in Western Europe
    http://www.stetson.edu/departments/religion/lucas/content/
    interior_art/interior_art_thumbnail.shtml
    This site contains a number of examples of ancient art, including cave
    paintings.
- Leonardo da Vinci Museum
    http://www.leonardo.net/gallery.html#start
    This site contains a number of example sketches by da Vinci.
- Leonardo da Vinci Posters and Prints
    http://www.allposters.com/gallery.asp?aid=473830&c=c&search=1526
    &cat=1526
    Example art work in poster and print form.
- U.S. Patent Office Search Engine
    http://patft.uspto.gov/
    Search for patents from 1976 to the present using text terms in any field.
    Once you bring up an individual patent, click on the images link at the
    bottom and then the drawings link to the left to see the patent drawings.

Books and Other Publications

- Bertoline, G. R. & Wiebe, E. N. (2004). *Technical Graphics Communication*. New York: McGraw-Hill.
    This book provides detailed background on engineering and technical graphics.
- Keller, P. R. & Keller, M. M. (1993). *Visual Cues*. Piscataway, NJ: IEEE Press.
    This book has lots of examples of advanced visualization techniques.
- Kosslyn, S. M. (1994). *Elements of Graph Design*. San Francisco, CA: W. H. Freeman.
    This book provides some guidelines for creating charts and graphs.

- Tufte, E. R. (1983). *The Visual Display of Quantitative Information.* Cheshire, CT: Graphics Press.
  A classic book on the presentation of data-driven graphics.

Project 2: Data-Driven Graphics: Graphing Maximum and Minimum Temperatures

Web Sites

- Climatological Normals Tables
  http://ols.nndc.noaa.gov/plolstore/plsql/olstore.prodspecific?prodnum=C00095-PUB-A0001#TABLES
  Source of the data used in this project. Click on the data table of interest.
- NCDC: US Climate at a Glance
  http://lwf.ncdc.noaa.gov/oa/climate/research/cag3/cag3.html
  Climate trends for select U.S. cites/states/regions. Graphs and data. An alternate source of temperature data. Choose Cities and enter information into the form. Choose table output.
- National Oceanic and Atmospheric Administration
  http://www.noaa.gov/
  Home page for NOAA. Responsible for weather satellites and other weather-related data collection.
- NOAA National Data Center
  http://nndc.noaa.gov/
  Home page for national weather data.

Project 3: Data Driven Graphics: Graphing Degree-Day Data

Web Sites

- Climatological Normals Tables
  http://ols.nndc.noaa.gov/plolstore/plsql/olstore.prodspecific?prodnum=C00095-PUB-A0001#TABLES
  Source of the data used in this project. Click on the data table of interest.
- NCDC: US Climate at a Glance
  http://lwf.ncdc.noaa.gov/oa/climate/research/cag3/cag3.html
  Climate trends for select U.S. cites/states/regions. Graphs and data. An alternate source of temperature data. Choose Cities and enter information into the form. Choose table output.
- National Oceanic and Atmospheric Administration
  http://www.noaa.gov/
  Home page for NOAA. Responsible for weather satellites and other weather-related data collection.
- NOAA National Climatic Data Center
  http://ncdc.noaa.gov
  Home page for national weather data.

Project 4: Conceptual Graphics: The Value of Insulation

Web Sites

- Home Heating Energy Calculations
  http://hyperphysics.phy-astr.gsu.edu/hbase/thermo/heatloss.html
  Source for theory and equations for calculating heat loss rate.

- U.S. Department of Energy, Insulation Fact Sheet
    http://www.ornl.gov/sci/roofs+walls/insulation/ins_01.html
    Background information on insulation technology and use.
- Florida Power & Light, Home Energy Advisor
    http://www.fpl.com/home/energy_advisor/contents/index.shtml
    Links to home energy basics and insulation use.
- DOE Publication on Residential Gas Prices
    http://www.eia.doe.gov/oil_gas/natural_gas/analysis_publications/
    natbro/gasprices.htm
    Provides breakdown of what goes into the cost of natural gas, along
    with average prices for different parts of the country.

Project 5: Animating Insulation Principles

Web Sites

- Home Heating Energy Calculations
    http://hyperphysics.phy-astr.gsu.edu/hbase/thermo/heatloss.html
    Source for theory and equations for calculating heat loss rate.
- U.S. Department of Energy, Insulation Fact Sheet
    http://www.ornl.gov/sci/roofs+walls/insulation/ins_01.html
    Background information on insulation technology and use.
- Florida Power & Light, Home Energy Advisor
    http://www.fpl.com/home/energy_advisor/contents/index.shtml
    Links to home energy basics and insulation use.
- DOE Publication on Residential Gas Prices
    http://www.eia.doe.gov/oil_gas/natural_gas/analysis_publications/
    natbro/gasprices.htm
    Provides breakdown of what goes into the cost of natural gas, along
    with average prices for different parts of the country.

Project 6: Multimedia Presentation of Insulation Properties

Web Sites

- Home Heating Energy Calculations
    http://hyperphysics.phy-astr.gsu.edu/hbase/thermo/heatloss.html
    Source for theory and equations for calculating heat loss rate.
- U.S. Department of Energy, Insulation Fact Sheet
    http://www.ornl.gov/sci/roofs+walls/insulation/ins_01.html
    Background information on insulation technology and use.
- Florida Power & Light, Home Energy Advisor
    http://www.fpl.com/home/energy_advisor/contents/index.shtml
    Links to home energy basics and insulation use.
- DOE Publication on Residential Gas Prices
    http://www.eia.doe.gov/oil_gas/natural_gas/analysis_publications/
    natbro/gasprices.htm
    Provides breakdown of what goes into the cost of natural gas, along
    with average prices for different parts of the country.

UNIT 2
3D Modeling

Project 1: Pump Technology
Web Sites

- Pump House
  http://pumpsweb.com/pumphouse/index.php
  Different types of pumps you can purchase are listed.
- Screw Pumps
  http://ewr.cee.vt.edu/environmental/teach/wtprimer/screw/screw.html
  This site provides background information on screw pumps.
- Metering Fluids in Medical Devices
  http://www.knf.com/magmetering.htm
  This site provides an in-depth discussion on the strengths and weaknesses of different types of medical pumps.
- Peristaltic Pumps
  http://www.williamsonpumps.co.uk/PERIBASICS.html
  This site offers background information on peristaltic pump technology.
- Periflo Pumps
  http://www.periflo.com/
  Industrial (non-medical) peristaltic hose pumps are highlighted.
- Anko Products
  http://ankoproducts.com/
  Various types of pumps are presented.
- Internet Glossary of Pumps
  http://www.animatedsoftware.com/pumpglos/peristal.htm
  A simple 2D animation of a peristaltic pump is featured.
- History of Insulin Pumps
  http://www.picturetrail.com/gallery/view?p=999&gid=201986&uid=103405
  This site provides a graphic history of the insulin pump.

Project 2: Part Modeling
Web Sites

- Tutorial Outpost
  http://www.tutorialoutpost.com/
  Tutorials for a range of 2D and 3D graphic software packages, specifically 3D Studio MAX.
- 3D Studio MAX 3.1 Tutorials
  http://www.cglearn.com/tutorials/max3/
  This site contains 3D Studio MAX 3.1 tutorials by Aaron Ross.
- 3D Artists: Tutorials—3D Studio MAX
  http://www.raph.com/3dartists/tutorials/t-3dsmax.html
  Tutorials for a range of 3D modeling software packages are presented, including 3D Studio MAX.

- 3D Studio MAX Tutorials
    http://www.tutorialized.com/tutorials/3DS-MAX/1
    This site contains tutorials and discussion forums for a range of 2D
    and 3D graphic software packages, including 3D Studio MAX.
- 3DLinks.com: Ultimate 3D Links—Tutorials: 3ds Max
    http://www.3dlinks.com/links.cfm?categoryid=13&subcategoryid=105
    Tutorials and discussion forums for 3D graphic software packages are
    presented, including 3D Studio MAX.
- 3D Studio MAX—3D Studio MAX Tutorials
    http://www.huntfor.com/3d/tutorials.htm
    This site includes 3D Studio Max tutorials both as demonstrations and
    for purchase.

Project 3: Assembly Modeling

Web Sites

- Tutorial Outpost
    http://www.tutorialoutpost.com/
    Tutorials for a range of 2D and 3D graphic software packages, specifi-
    cally 3D Studio MAX.
- 3D Studio MAX 3.1 Tutorials
    http://www.cglearn.com/tutorials/max3/
    This site contains 3D Studio MAX 3.1 tutorials by Aaron Ross.
- 3D Artists: Tutorials—3D Studio MAX
    http://www.raph.com/3dartists/tutorials/t-3dsmax.html
    Tutorials for a range of 3D modeling software packages are presented,
    including 3D Studio MAX.
- 3D Studio MAX Tutorials
    http://www.tutorialized.com/tutorials/3DS-MAX/1
    This site contains tutorials and discussion forums for a range of 2D
    and 3D graphic software packages, including 3D Studio MAX.
- 3DLinks.com—Ultimate 3D Links - Tutorials: 3ds Max:
    http://www.3dlinks.com/links.cfm?categoryid=13&subcategoryid=105
    Tutorials and discussion forums for 3D graphic software packages are
    presented, including 3D Studio MAX.
- 3D Studio MAX—3D Studio MAX Tutorials
    http://www.huntfor.com/3d/tutorials.htm
    This site includes 3D Studio Max tutorials both as demonstrations and
    for purchase.

Project 4: Animation

Web Sites

- Tutorial Outpost
    http://www.tutorialoutpost.com/
    Tutorials for a range of 2D and 3D graphic software packages are pre-
    sented, specifically 3D Studio MAX.
- 3D Studio MAX 3.1 Tutorials
    http://www.cglearn.com/tutorials/max3/
    This site contains 3D Studio MAX 3.1 tutorials by Aaron Ross.
- 3D Artists: Tutorials—3D Studio MAX
    http://www.raph.com/3dartists/tutorials/t-3dsmax.html
    Tutorials for a range of 3D modeling software packages are presented,
    including 3D Studio MAX.

- 3D Studio MAX Tutorials

    http://www.tutorialized.com/tutorials/3DS-MAX/1

    This site contains tutorials and discussion forums for a range of 2D and 3D graphic software packages, including 3D Studio MAX.

- 3DLinks.com: Ultimate 3D Links—Tutorials: 3ds Max

    http://www.3dlinks.com/links.cfm?categoryid=13&subcategoryid=105

    Tutorials and discussion forums for 3D graphic software packages are presented, including 3D Studio MAX.

- 3D Studio MAX—3D Studio MAX Tutorials

    http://www.huntfor.com/3d/tutorials.htm

    This site includes 3D Studio Max tutorials both as demonstrations and for purchase.

Project 5: Rendering and Deformation

Web Sites

- Tutorial Outpost

    http://www.tutorialoutpost.com/

    Tutorials for a range of 2D and 3D graphic software packages are included, specifically 3D Studio MAX.

- 3D Studio MAX 3.1 Tutorials

    http://www.cglearn.com/tutorials/max3/

    This site contains 3D Studio MAX 3.1 tutorials by Aaron Ross.

- 3D Artists: Tutorials—3D Studio MAX

    http://www.raph.com/3dartists/tutorials/t-3dsmax.html

    Tutorials for a range of 3D modeling software packages are presented, including 3D Studio MAX.

- 3D Studio MAX Tutorials

    http://www.tutorialized.com/tutorials/3DS-MAX/1

    Tutorials and discussion forums for a range of 2D and 3D graphic software packages are presented, including 3D Studio MAX.

- 3DLinks.com: Ultimate 3D Links—Tutorials: 3ds Max

    http://www.3dlinks.com/links.cfm?categoryid=13&subcategoryid=105

    This site contains tutorials and discussion forums for 3D graphic software packages, including 3D Studio MAX.

- 3D Studio MAX—3D Studio MAX Tutorials

    http://www.huntfor.com/3d/tutorials.htm

    This site contains 3D Studio MAX tutorials both as demonstrations and for purchase.

Project 6: Camera Viewpoints and Particle Systems

Web Sites

- Glossary of Optical Terms

    http://www.dpreview.com/learn/?/Glossary/Optical/

    This site contains definitions of terms used to specify camera movements and viewpoints.

- Tutorial Outpost

    http://www.tutorialoutpost.com/

    Tutorials for a range of 2D and 3D graphic software packages are presented, including 3D Studio MAX.

- 3D Studio MAX 3.1 Tutorials
    http://www.cglearn.com/tutorials/max3/
    This site contains 3D Studio MAX 3.1 tutorials by Aaron Ross.
- 3D Artists: Tutorials—3D Studio MAX
    http://www.raph.com/3dartists/tutorials/t-3dsmax.html
    Tutorials for a range of 3D modeling software packages are presented, including 3D Studio MAX.
- 3D Studio MAX Tutorials
    http://www.tutorialized.com/tutorials/3DS-MAX/1
    Tutorials and discussion forums for a range of 2D and 3D graphic software packages are included, specifically 3D Studio MAX.
- 3DLinks.com: Ultimate 3D Links—Tutorials: 3ds Max
    http://www.3dlinks.com/links.cfm?categoryid=13&subcategoryid=105
    Tutorials and discussion forums for 3D graphic software packages are presented, including 3D Studio MAX.
- 3D Studio MAX—3D Studio MAX Tutorials
    http://www.huntfor.com/3d/tutorials.htm
    This site includes 3D Studio MAX tutorials, both as demonstrations and for purchase.

UNIT 3
Energy and Power Technology

Project 1: History of Energy and Power Technologies

Web Sites

- Comparing Alternative Energy Forms
    http://library.thinkquest.org/17658/
    A ThinkQuest developed by high school students that focuses on learning about and comparing alternative energy forms.
- National Renewable Energy Laboratory: Clean Energy Basics
    http://www.nrel.gov/clean_energy/
    This Web site developed by the National Renewable Energy Laboratory focuses on many renewable energy resources being researched by the U.S. Department of Energy. Some of the energy sources discussed include solar energy, wind energy, biomass energy, hydrogen energy, geothermal energy, hydroelectric power, and ocean energy. The site also discusses environmental and economic impacts of using renewable energy sources. There is a page of Web resources specifically for students and teachers to use in learning about each type of energy source.
- U.S. Department of Energy's KIDS Zone
    http://www.energy.gov/engine/content.do?BT_CODE=KIDS
    This site developed by the U.S. Department of Energy is written for students, and it is about energy. Useful links include Energy Basics and History and Milestones of Energy, which includes a timeline of important energy events.

Project 2: Technological Tools and Energy

Web Sites

- Classroom Energy
  http://www.classroom-energy.com
  A Web site developed by the American Petroleum Institute discussing the petroleum and natural gas industries. The Web site is designed for use in the classroom and includes some multimedia presentations that illustrate the new technologies being developed and used in the production of petroleum and natural gas products. Be sure to request a free CD version of the Web site for use in your classroom.

- Comparing Alternative Energy Forms
  http://library.thinkquest.org/17658/
  A ThinkQuest developed by high school students that focuses on learning about and comparing alternative energy forms.

- Energy Is Everywhere
  http://www.energymatch.com/features/artkids.asp?articleid=39
  A series of articles on many energy topics including power plants, life before electricity, conservation methods, and renewable versus non-renewable resources.

- Energy Quest—Energy Education from the California Energy Commission
  http://www.energyquest.ca.gov/index.html
  This Web site has many useful links to a variety of resources that will increase students' understanding of energy concepts. Links that may be especially useful include the energy conversion calculator under The Energy Library, The Energy Story, and Super Scientists.

- Howstuffworks "Science Channel"
  http://science.howstuffworks.com/
  From this Web site, connect to links that explain how batteries, solar cells, electric cars, and fuel cells work. The pages on batteries also include information on how to build your own batteries from common materials.

- National Energy Education Development (NEED) Project
  http://www.need.org
  This site describes national energy education projects and programs available for use in schools. Curricular materials have been developed and may be accessed from the site for all grade levels. There are also many links to other energy education resources.

- Photovoltaics in a High School Lab
  http://www.botproductions.com/pv/pv.html
  Three high school sophomore students developed this site as part of a project in their chemistry class. For their project, they researched photovoltaic cells and actually built 12 working cells in their high school science class. Information is given on how to build your own photovoltaic cells.

- Rayovac's World of Batteries
  http://www.rayovac.com/worldofbatteries/
  This site has many educational links relating to the production, use, and disposal of batteries. In particular, teachers and students may find the How a Battery Is Made, How a Battery Works, and Cool Activities links especially useful. The Cool Activities link has a worksheet for determining the number of batteries disposed of annually in household garbage.

- U.S. Department of Energy PV Home Page
  http://www.eere.energy.gov/solar/photovoltaics.html
  This Web site developed by the U.S. Department of Energy has a multitude of resources on how photovoltaic cells are made and work. There is an animation of how a solar cell works and links to online resources to use in classrooms when learning about solar energy.

Project 3: Visualizing Energy Transfer Devices

Web Sites

- Amazing Rube Goldberg and the Invention Convention
  http://www.rube.iscool.net/
  Learn more about Rube Goldberg and the Rube Goldberg competition sponsored by the Invention Convention.
- Energy: Fuel for Thought (Bureau of Land Management—Environmental Education)
  http://www.blm.gov/education/00_resources/articles/energy/index.html
  This Web site was developed by the Bureau of Land Management regarding various energy resources. There are many educational activities suggested, including a downloadable poster of a fictional middle school in which you identify the various forms of energy, energy sources, and the energy transformations that are taking place.

Project 4: The Environmental Impacts of Batteries

Web Sites

- Duracell.com—Battery Disposal Center
  http://www.duracell.com/care_disposal/disposal.asp
  This site provides an overview of how to dispose of different types of batteries properly.
- Energizer.com—Learning Center
  http://www.energizer.com/learning/default.asp
  This site has many educational links relating to the production, use, and disposal of batteries. In particular, teachers and students may find the How a Battery Works link especially useful.
- Howstuffworks "Science Channel"
  http://science.howstuffworks.com/
  From this Web site, find links to material that explains how batteries, solar cells, electric cars, and fuel cells work. The pages on batteries also include information on how to build your own batteries from common materials.
- Lead.org—Lead Acid Battery Problems
  http://www.lead.org.au/lanv3n2/lanv3n2-5.html
  This site presents information on the harmful effects batteries can have.
- Minnesota Pollution Control Agency—Managing Spent Lead Acid Battery Casings at Residential Sites.
  http://www.pca.state.mn.us/publications/w-hw4-09.pdf
  This article provides a lot of information about what happens when lead acid batteries are found buried in people's backyards.

- Photovoltaics in a High School Lab
  http://www.botproductions.com/pv/pv.html
  Three high school sophomores developed this site as part of a project in their chemistry class. For their project, they researched photovoltaic cells and actually built 12 working cells in their high school science class. Information is given on how to build your own photovoltaic cells.
- Rayovac's World of Batteries
  http://www.rayovac.com/worldofbatteries/
  This site has many educational links relating to the production, use, and disposal of batteries. In particular, teachers and students may find the How a Battery Is Made, How a Battery Works, and Cool Activities links especially useful. The Cool Activities link has a worksheet for determining the number of batteries disposed of annually in household garbage.

Project 5: Fuel Cell Technology and the Hydrogen Economy

Web Sites

- Project Options
  http://www.fwee.org/TG/nwaterpwr.html
  The Foundation for Water and Energy Education: The Nature of Water Power curriculum engages students in "hands-on" investigations of hydropower.
- Project Extensions
  http://www.fwee.org/Tours.html
  The Foundation for Water and Energy Education's "Hydro Tours" Web site is an interactive site that allows students to click on, view images of, and learn about the various components of a typical hydropower unit.
  http://www.fwee.org/TG/wpmodel.html
  Make your own turbine with help from the Foundation for Water and Energy Education's Web site.
  http://www.eere.energy.gov/windandhydro/
  The U.S. Department of Energy's Hydropower Technologies Program Web site contains information about work being done in advanced turbine development and hydropower integration with other renewable energy sources.
- Howstuffworks "Science Channel"
  http://science.howstuffworks.com/
  From this Web site, find links to material that explains how batteries, solar cells, electric cars, and fuel cells work.
- Comparing Alternative Energy Forms
  http://library.thinkquest.org/17658/
  A ThinkQuest developed by high school students that focuses on learning about and comparing alternative energy forms.
- National Renewable Energy Laboratory: Clean Energy Basics
  http://www.nrel.gov/clean_energy/
  This Web site developed by the National Renewable Energy Laboratory focuses on many renewable energy resources being researched by the U.S. Department of Energy. Some of the energy sources discussed include solar energy, wind energy, biomass energy, hydrogen energy, geothermal energy, hydroelectric power, and ocean energy. The site also discusses the environmental and economic impacts of using renewable energy sources. There is a page of Web resources specifically for students and teachers to use in learning about each type of energy source.

Project 6: Small-scale Hydroelectric Power

Web Sites

- History
  http://www.eere.energy.gov/windandhydro/hydro_history.html
  The U.S. Department of Energy's (DOE) Web site highlights some key
  dates in the history and use of alternative power.
  http://www.teslasociety.com/
  The Web site for the Tesla Memorial Society of New York contains a
  wealth of information about the lifework of Nikola Tesla.
- Current Statistics
  http://www.eia.doe.gov/kids/infocardnew.html#ELECTRICITY
  Part of the U.S. Department of Energy's Energy for Kids Web site,
  which contains usage statistics for most types of energy including
  hydropower.
  http://www.eia.doe.gov/cneaf/solar.renewables/page/trends/table4.html
  The Energy Information Administration publishes official energy
  statistics from the U.S. government.
- Turbines
  http://www.eere.energy.gov/windandhydro/hydro_turbine_types.html
  This site, which is maintained by the Department of Energy, describes
  the various types of hydropower turbines that exist.
  http://www.canren.gc.ca/tech_appl/index.asp?CaId=4&PgID=1154
  The Canadian Renewable Energy Network's (CanREN) Web site
  explains the different types of turbines.
- Kilowatts & Megawatts
  http://www.uwsp.edu/cnr/wcee/keep/Mod1/Whatis/energymeasures.htm
  Learn how energy is measured and quantified by using this Web resource.
  http://michaelbluejay.com/electricity/cost.html#kilowatt
  Visit the self-proclaimed Mr. Energy's Web site and find the answers
  to questions such as: What the heck is a kilowatt hour? and How do I
  find out how much electricity something uses?
  http://www.nationmaster.com/graph-T/ene_ele_con
  NationMaster.com is a massive central data source and a handy way to
  graphically compare nations on such things as energy consumption.
- Locating a Source
  http://water.usgs.gov/nsip/nsipmaps/federalgoals2.html
  The U.S. Geological Survey (USGS) maintains The National Streamflow Information Program. Here you can access interactive Streamgage maps and data for your state.
- Maps
  http://www.usgs.gov/
  You can use U.S. Geological Survey maps of your area to gather elevation data during the assessment of your source's head.
- Equipment Costs and Availability
  http://microhydropower.net/directory/manufacturers.php
  Microhydropower's business directory, including a list of manufacturers and links to their Web sites.
  http://www.backwoodssolar.com/ under "Our Online Catalog," select
  "Hydropower"
  This site contains an on-line catalog featuring planning aids, turbines,
  and generators for microhydropower system development.

- Local Requirements and Restrictions
  http://www.naseo.org/
  The National Association of State Energy Officials (NASEO) site provides current contact information for state energy offices, including links to their Web sites.
  http://www.ferc.gov/industries/hydropower.asp
  The Federal Energy Regulatory Commission (FERC) provides information about federal requirements and restrictions that may impact the development of hydropower at your site.
  http://www.usace.army.mil/
  U.S. Army Corps of Engineers.
- Risks and Benefits of Hydroelectric Power
  http://www.canren.gc.ca/tech_appl/index.asp?CaId=4&PgId=43
  The Canadian Renewable Energy Network (CanREN) Web site outlines how building a hydroelectric system impacts the environment and describes measures that can be taken to lessen or avoid any negative effects of this technology.
- Preliminary Feasibility Assessment Software
  http://www.retscreen.net/ang/g_small.php
  RETScreen International Clean Energy Project Analysis Software.
- Common Conversion Factors
  http://www.onlineconversion.com/flow_rate_volume.htm
  An energy conversion calculator

UNIT 4
Nanotechnology

Project 1: Overview of the Field and the Nanoscale

Web Sites

- Powers of Ten
  http://www.powersof10.com/
  Explores the relative size of object
- Molecular Expressions
  http://micro.magnet.fsu.edu/
  Explores the world of optical microscopy
- National Nanotechnology Initiative
  http://www.nano.gov/
  Research to further the understanding of nanoscale phenomena

Project 2: What Is Nanoscience?

Web Sites

- Atoms and Molecules
  http://www.chemistry.mcmaster.ca/bader/aim/
  An introduction to the electronic structure of atoms

Project 3: Tools of Nanotechnology

Web Sites

- Center for Responsible Nanotechnology
    http://crnano.org/overview.htm
    Current findings in molecular manufacturing
- NM Interdisciplinary Education Group
    http://www.mrsec.wisc.edu/Edetc/LEGO/bookindex.html
    Using Lego™ to explore nanotechnology concepts
- NanoHUB
    http://www.nanohub.org/news/nuggets/Childrens_Museum_Exhibit_in_Oak_Ridge/
    An example of how a museum project used Lego™ in a nano exhibit
- Molecular Assembler Debate
    http://pubs.acs.org/cen/coverstory/8148/8148counterpoint.html
    Debate between K. Eric Drexler and Richard E. Smalley on the viability of molecular assemblers
- Scanning Probe Microscopy Techniques
    http://www.nanoscience.com/education/tech-overview.html
    Techniques used by nanoscientists with STM and AFM technologies

Project 4: Ethical and Societal Concerns Related to Nanotechnology

Web Sites

- Center on Nanotechnology and Society
    http://www.nano-and-society.org/news/
    Ethical, legal, and societal issues in nanotechnology
- The A to Z of Nanotechnology
    http://www.azonano.com/news.asp?newsID=1284
    Nanotechnology-related current events
- Center for Responsible Nanotechnology
    http://crnano.org/safe.htm
    Articles about the societal impact and measuring risks of nanotechnology
- Foresight Nanotech Institute
    http://www.foresight.org/
    Institute dedicated to beneficial uses of nanotechnology

Survey Design Information
- Creative Research Systems
    http://www.surveysystem.com/sdesign.htm
    A for-profit research firm with an online survey design tutorial

Project 5: Forces and Effects at the Nanoscale

Web Sites

- NOVA Science in the News
    http://www.science.org.au/nova/077/077key.htm
    Revolutionary breakthroughs in nanoscience

- Nanotechnology Now
  http://www.nanotech-now.com/interviews.htm
  Interviews with nanotechnology researchers, educators, and entrepreneurs
- New Scientist
  http://www.newscientist.com/article.ns?id=dn4406
  Article on electronic components made with biological self-assembly
- Wikipedia
  http://en.wikipedia.org/wiki/Self-assembly
  Entry on biological self-assembly
- International Society of Complexity, Information and Design
  http://www.iscid.org/encyclopedia/Molecular_Machines
  Molecular machines
- Nanoscale Properties
  http://nanosense.org/activities/sizematters/index.html
  Background material on nanoscale, including forces at that scale

Project 6: Applications in the Field of Nanotechnology

Web Sites

- Institute for Nanotechnology
  http://www.nano.org.uk/
  Discussion forums on nanotechnology
- Current Applications of Nanotechnology
  http://www.nanotech-now.com/current-uses.htm
  Current nanotechnology applications
- Nanotechnology Now
  http://www.nanotech-now.com/news.cgi?story_id=07159
  Rapid progression in molecular electronics

UNIT 5
Biometrics

Project 1: Biometric Tools

Web Sites

- National Biometric Security Project
  http://www.nationalbiometric.org/about_history.php
  A timeline of milestones in biometrics
- Newsfactor Magazine
  http://www.newsfactor.com/news/Improving-E-Commerce-withBiometrics/story.xhtml?story_id=0010002ECQV4
  Improving E-commerce security with biometrics
- Biometrics Research
  http://biometrics.cse.msu.edu/
  Biometrics research at Michigan State University

Project 2: Collecting and Analyzing Fingerprints

Web Sites

- Ridges and Furrows
    http://www.ridgesandfurrows.homestead.com/fingerprint_patterns.html
    Identifies fingerprint patterns based on the Henry Classification System
- Cyberbee Fingerprinting
    http://www.cyberbee.com/whodunnit/fp.html
    Taking, classifying, and dusting for fingerprints
- Crimescene Detective Store
    http://www.crimescene.com/store/index.php?main_page=index/
    Source of latent fingerprint kits and other forensic supplies
- Lynn Peavey Company
    http://www.lynnpeavey.com/
    Source of latent fingerprint kits and other forensic supplies

Project 3: Processing Fingerprints

Web Sites

- Free Image Processing and Analysis Software
    http://www.ridgesandfurrows.homestead.com/downloads.html
    Serves as an introduction to digital image enhancement

Project 4: Identity Theft and Privacy

Web Sites

- Biometrics and Security
    http://www.infosyssec.net/infosyssec/security/biomet1.htm
    Provides an overview of biometrics as a form of security
- The Thin Blue Line
    http://www.policensw.com/info/fingerprints/finger07.html
    Discusses the anatomy of fingerprints

Project 5: Crime Scene Visualization

Web Sites

- Visualization Toolkit
    http://public.kitware.com/VTK/
    Open source, freely available software system for 3D computer graphics, image processing, and visualization
- UM Supercomputing Institute
    http://www.msi.umn.edu/user_support/software/Visualization.html
    Listing of visualization software
- Graphics, Animation, and Visualization
    http://www.tenlinks.com/CAD/products/graphics.htm
    2D and 3D visualization software
- Design Ware
    http://www.designwareinc.com/3d_prod.htm
    3D crime scene and accident reconstruction visualization software

- The ARC Network
  http://www.accidentreconstruction.com/products/accident-reconstruction-software.asp
- North East Multi Regional Training NEMRT
  http://user.mc.net/~paulb/ck97class/
  3D crime scene and accident reconstruction visualization using CADKEY97 software

Project 6: Biometric Device Design

Web Sites

- ZVETCO biometrics
  http://www.zvetcobiometrics.com/
  Supplier of network security devices and applications
- Integrated Solutions
  http://www.integratedsolutionsmag.com/index.php?option=com_jambozine&layout=article&view=page&aid=4915&Itemid=5
  Explores biometric security options for the workplace

Glossary

3D Cartesian coordinate system   This system is composed of three points: X, Y, and Z. A positive X specifies points to the right, and a positive Y is up, a positive Z is toward the viewer. Together X, Y, and Z create a vector that can identify any point within a three dimensional space.

absolute coordinate system   Also referred to as the absolute reference frame. This coordinate system ultimately identifies a complete scene within the virtual world through the 3D Cartesian coordinate system.

absolute size   The magnitude of a single number in comparison to the size of another number or series of numbers.

algorithms   Step-by-step procedures for solving a problem.

alkaline batteries   A type of nonrechargeable battery often used in consumer equipment.

alternating current (AC)   Electric current that reverses direction many times per second.

alphabet   A set of letters, characters, or symbols in which one or more languages are written.

ambient light   The surrounding the light of an object in a scene. Provides an overall level of light for the entire scene.

angle of view   The segment of an object that is seen by the viewer as determined by the focal length.

animation   A graphic technique used to depict change over time through a rapidly projected series of images. The rapid sequencing produces a psychological effect called stroboscopic motion and produces the appearance of real motion. Animation can be created using manual or computer-based tools.

anode   Positive electrode where electrons are lost or oxidized in oxidation–reduction (redox) reactions.

aperture   The size of the hole that lets light into the camera. The larger the aperture the shallower the depth of field.

Archimedes pump   The Archimedes screw pump, commonly referred to as the snail pump, was used to supply water for irrigation systems

array   Copying multiple parts simultaneously is considered an array operation.

artifacts   In an identity theft that uses artifacts, the thief presents a replication of the biometric such as a mold of a finger with a fingerprint impression from the victim.

assembly   The act of placing previously fabricated and refined components together into a finished piece.

ASCII (American Standard Code for Information Interchange)   Commonly used by computer systems for encoding text information in a common, binary format that all computer systems can understand. It does not contain any formatting information such as bolding or italicizing of characters. Certain special characters, though, they can be typed from a keyboard, may not be part of the standard ASCII character set. Often referred to as "raw text" or just "text" format.

asbestos   Tiny, fibrous minerals found naturally throughout much of the earth. Used in insulation, flooring, and roofing.

atom   The smallest unit of a chemical element.

atomic force microscope (AFM)   A force microscope used in nanotechnology that works by measuring a magnetic or mechanical force on the surface of the sample.

atomic scale   The scale of working with individual atoms. This is the scale nanotechnology works at.

audience   A group of listeners, viewers, or spectators. This term is used to define who is intended to receive and understand the multimedia information you have produced.

authentication   Process of making a match between a biometric trait and another piece of information. Also known as verification.

Automatic Fingerprint Authentication Systems (AFAS)   The input is an ID plus a fingerprint. The output is either a "yes, you are who you say you are," or "no, you're not who you say you are."

Automatic Fingerprint Identification Systems (AFIS)   Historically used by law enforcement agencies and common in crime scene investigations. The input is the fingerprint, and the output is a match with the person to whom the fingerprint belongs.

axial   Axial flow devices carry fluid in the same direction as the rotating axis. In a closed chamber, the pressure and flow of the fluid can be controlled.

base part   This part is considered a core part of an object, around which other parts are attached.

battery   A device that stores energy and uses an electrochemical reaction to provide electrical energy.

bilge pumps   This type of pump is used to pump excess water from inside of boats. Bilge pumps use diaphragm pumps with leather membranes.

biomass   Plant material, vegetation, or agricultural waste available on a renewable basis. Biomass created from inexpensive plant sources can be converted to fuels (hydrogen, ethanol and diesel), pharmaceutical, and other chemicals for the manufacturing sector through chemical processing.

biometric   A specific physical or behavioral characteristic.

biometrics   The science and technology involved in collecting and analyzing data about physical or behavioral individual traits used for identifying or verifying the identity of a person.

biometrics device   A device that works by analyzing a biometric that can identify an individual and comparing it to a database or library containing information on that biometric.

blend   Also referred to as an extrusion, a blend is creating a three-dimensional object from a two-dimensional outline by forcing material outward to obtain length and give height.

Boolean operations   Operation in which a separate 3D form is temporarily created that either overlaps or adjoins the existing 3D model. This new form is either added or subtracted from the existing model.

brightness   The level of light energy. In terms of color definition, it indicates how "light" or "dark" a color is.

British thermal unit (Btu)   A non-metric unit of energy, now primarily used in the United States. A Btu is defined as the amount of heat energy required to raise one pound of water one degree Fahrenheit. Btu is often used to describe the heat value of fuels or the heating or cooling capacity of a system.

bottom-up approach   A nanotechnology technique that investigates ways to build and manufacture structures molecule by molecule or atom by atom.

building envelope   The border between the conditioned (heated or cooled) space and the exterior unconditioned space. The envelope can often be thought of as a continuous 3D surface that separates the inside from the outside. The goal is usually to have an envelope that minimizes the movement of heat energy from one side to the other.

bump mapping   With this technique the user can manipulate how light is reflected off surfaces. Rather than gradually changing light reflection across a surface, bump mapping allows regular small changes in light reflection happen across the surface.

cam   A component part of the finger pump that is used to press down a series of fingers onto a bladder or tube. This can be done in one pass on a disposable bladder or as continuous motion on a tube.

carbon nanotubes   Carbon tubes at the nanoscale that have strength by weight that is 600 times that of steel. If feasible to manufacture, they would be useful in military defense, and building materials.

cathode   Negative electrode where electrons are gained or reduced in oxidation–reduction (redox) reactions.

cell-based animation   A type of animation in which each cell, or frame, of the animation is individually rendered. Each cell is slightly different from the previous, giving the appearance of change when they are projected in series. This is the oldest, traditional form of animation.

chemical energy   A form of energy that is released as a result of chemical reactions.

children   A single part that is added to a parent part during assembly.

circular sweep   A circular sweep is done about an axis and can sweep a profile any number of degrees about this axis. Different forms can be created by locating the sweeping axis on or off the edge of the profile.

climatic   Of, or relating to, climate. Climate is a description of weather conditions, such as temperature or rainfall, usually over a period of years.

climatological normals   The average weather measurement over a specified period of time for a particular location. An example would be the average monthly temperature in Chicago.

coding   To transform information into an alternate form by means of a systematic transformation. An example would be to code numeric values, such as temperature, with specific colors.

color   A visual element of objects that is a direct outcome of light production, diffusion, and reflection.

composite board   A building product in sheet or board form made from finely ground wood chips held together with a resin. Sometimes also referred to as particleboard or strandboard.

compound   A substance made up of atoms from at least two elements.

concept-driven graphic   A type of graphic visualization that codes ideas, or concepts, rather than numeric values as graphic elements. An example might be a diagram that depicts how a hurricane is formed. *See also* data-driven graphic.

conduction The transfer of heat or electrical energy by means of direct contact. Denser materials and metals in general are good conductors. *See also* convection and radiation.

cone An extruded circular shape that tapers to a point on one end.

contact mode In an atomic force microscope, the tip maintains a constant force while scanning over the surface.

convection The transfer of heat energy within a fluid or gas. The temperature difference within the fluid or gas often causes movement and, therefore, the distribution of the energy. *See also* conduction and radiation.

covalent bonds Atoms sharing electrons with one another.

crane Any camera shot that takes the user off the ground and films the scene at a height above a typical human view.

cuneiform An alphabet consisting of wedge-shaped characters. A system of writing common in the ancient Near East.

curved A continuous object with a geometrically rounded bend.

cylinder An extruded circular or elliptical shape.

data-driven graphic A type of graphic visualization that systematically codes numeric values or regular categories as graphic symbols. Traditional charts and graphs are examples of data-driven graphics. *See also* concept-driven graphic.

deformation Occurs when geometry has been altered by the interaction with other geometry. It can also be achieved by the user manipulating local portions of the model geometry.

degree days A unit of measurement that provides a value for how cold or warm a location was relative to a commonly accepted standard temperature. For heating a house, you would calculate this by subtracting the average outdoor temperature for the day by the standard indoor temperature, say 65 degrees Fahrenheit. Degree days are usually added up over a period of a month or season.

degrees of freedom The number of possible translations and rotations a movable part can make.

dependent variable A variable dependent on the response of the model, or system, and not controlled by the experimenter. In design testing, the dependent variable is often a performance criteria. For example, the stopping distance for new car tire designs. *See also* independent variable.

depth of field Indicates how much of the scene is in focus. The shallower the depth of field, the less of the scene in front or behind the focal point stays in focus.

desaturated To lower the saturation value of a color, usually to the point the color appears gray (or black or white). *See also* saturation.

design problem An engineering or technical problem by means of the design process. This process entails a clear definition of what the problem to be solved is and the application of tools and problem-solving processes to come up with the best solution for the problem. This process often includes the creation of graphics to help the designer work through the problem or communicate possible solutions to others.

diagram A graphic design that typically represents a process or the component parts of a system. The components can be represented by a combination of graphics and text, spatially arranged relative to each other.

diffuse reflection When light reflects off the surface of an object in all directions.

dipole moment A measure of the level of separation of the charge in a molecule.

direct current (DC) Electric current which flows in one direction.

directional Light rays are parallel to each other and at the same intensity no matter how far away the object is.

diversion (system)   Sometimes called run-of-river, it is a facility that channels a portion of a river through a canal or penstock. It may not require the use of a dam.

dolly   A camera movement where the focal length is not altered, but the viewer is transported along the line of sight.

doping   The intentional introduction of a substance that alters the conductivity of a semiconductor.

drywall   Also known as wallboard or gypsum board, it is a paper-covered panel of gypsum used as an interior wall covering in most new residential homes. It can be easily cut with a saw or broken along score lines created with a knife. Drywall can be plastered, painted, or covered with wallpaper.

duplication   Accomplished by applying rigid transformations to copies of parts.

dynamic   Marked by continuous activity or change. A dynamic visualization would be continuously changing, such as an animation.

dynamo   A generator consisting of a coil (the armature) that rotates between the poles of an electromagnet (the field magnet) causing a current to flow in the armature.

electrical energy   A form of energy that is associated with movement of electrons through a substance.

electrical grid   The network of power lines used to deliver electricity to inhabited areas.

electricity   A form of energy that is released by the flow or accumulation of electrons.

electrodes   The surfaces where oxidation and reduction occur.

electromagnetic pumps   Manipulates the electromagnetic field. Electrical wiring wrapped around a pipe creates a magnetic field force in the direction that the fluid should move.

electrons   Negatively charged particles.

energy   The capacity or ability to do work.

engineering   A profession concerned with applying scientific knowledge to practical problems. Engineering has historically been divided up into disciplinary fields working on different problem areas. These disciplines include civil, mechanical, biomedical, and aerospace.

extrusion   Creating a three-dimensional object from a two-dimensional outline by forcing material outward to obtain length and give height.

face recognition systems   Biometric devices that measure facial characteristics by scanning a person's face using a digital video camera, creating a digital image, and measuring various facial structures.

fiberglass   Extremely fine fibers of glass matted together and used as an insulating material. Also refers to a number of composite, epoxy, and ceramic materials that are similarly used as insulating material.

fingerprint   Impressions or prints made from ridges and valleys on finger pads. Fingerprints are unique to each individual.

fingerprint scanners   Biometric devices that analyze fingertip patterns by scanning a finger or fingers using one of various technologies and creating a digital image.

finger pump   Forces fluid to move inside a tube by pushing a chain of rods in sequence. Finger pumps are often found in medical applications, since no part touches the fluid and therefore the fluid remains sterile.

fixed   When a part is securely placed against or fastened to a primary location.

flow   The amount of water passing a given point within a given period of time.

flowchart A specific type of diagram graphically outlining a series of actions that should happen over a period of time and their conditional relationship to each other.

focal angle Determines whether a small or large part of the model is lit up. The focal angle creates a cone of light with the central axis of the cone being the direction of the spotlight.

focal length Distance between the viewing plane and its point of central attention.

forebay The impoundment immediately above a dam or hydroelectric plant intake structure. The term is applicable to all types of hydroelectric developments. fossil fuels Formed from ancient decayed plants and animals. Examples include oil, coal, and natural gas.

fossil fuels Formed from ancient decayed plants and animals. Examples include oil, coal, and natural gas.

frame A singular composed still image that is generally sequenced together with other frames to compose a moving animation.

frustrum Section of a solid that lies between two similar planes, splitting the solid.

fuel cell Device that converts the chemical energy of hydrogen (or a hydrogen-containing mixture) and oxygen from the air directly into electricity without combustion.

fuel cell cars Cars that utilize fuel cells.

gear pump A gear pump moves the liquid between the meshing teeth of two gears. As the teeth come together, the liquid is trapped, compressed, and pushed out the other side.

generator The electrical equipment in power systems that converts mechanical energy to electrical energy.

geothermal energy A form of energy that is derived from the heat inside the earth.

graph One of many types of concept or data-driven graphics. The most common types of graphs are data-driven graphics where one variable is represented on an X-axis and another variable is represented on a perpendicular Y-axis. Data values are plotted to represent the relationship between these two variables.

graphician An individual who designs or creates graphics. Usually referring to those who create graphics that are used for conveying specific information, such as engineering specifications.

gravitational energy A form of potential energy that an object has due to its position.

hand geometry readers Biometric devices that measure overall hand shape, finger length, finger thickness, palm size, overall area, relationship between fingers, and other features by capturing a 3D view of the hand.

head Vertical change in elevation, expressed in either feet or meters, between the head water level and the tailwater level.

heat energy A form of energy related to the heat of an object.

heavy metals A popular term used to refer to metallic elements that are particularly dense. Some of the most common ones are copper, lead, mercury, and zinc, though this term can refer to all metallic elements with an atomic weight between 63 and 200. Also, this term is often used to refer to metallic elements that are poisonous to mammals in small quantities.

hierarchical trees Chain of parts in a fabrication and assembly order. The base part is considered at the top of the tree with all other parts added to the assembly as children of the base part.

hierarchy   A graded or ranked order. A hierarchical graph is a diagram that spatially depicts the ranked relationship of components to each other.

hieroglyphics   Pictorially-based alphabet used as the basis of a the written language used in ancient Egypt.

hue   The aspect of color represented by the specific mixture of wavelengths. Hue is the specific name for what people commonly use the word color for: red, purple, yellow.

hydroelectric power   Energy created by the flow of water.

hydrogen bond   Created when the negative oxygen end of a molecule aligns with the hydrogen end of a neighboring molecule. This creates the sharing of an electron with a hydrogen atom.

hydrogen energy   A form of energy that utilizes hydrogen for various purposes.

identification   Process of comparing a given trait to a database or library of characteristics for a large population. It is most often used by law enforcement agencies.

identity theft   Act of impersonating another person by using the victim's information.

imitations   In an identity theft that uses imitation, the thief impersonates another person.

impoundment   A confined body of water such as a reservoir or lake. Typically created by a dam to store water that is released to meet or maintain authorized purposes.

impressions   Fingerprints made by contact with a soft, pliable surface such as clay, wax, wet paint, or blood.

independent variable   A variable controlled by the experimenter when investigating the properties of a system or process. In design testing, the independent variable is often based on the different design variations you are exploring. For example, the overall weight of the rocket might be varied for different rocket designs and be the independent variable you control. See also dependent variable.

insulation   Either the material or the process of design used to slow the flow of thermal energy from one side of a barrier to the other. Common insulation materials include fiberglass, foam, and plastic sheeting.

intersection   The area of overlap is all that remains of the model after the operation.

ion   An atom or group of atoms with a positive or negative electrical charge.

ionic bonds   Atoms transferring electrons between one another.

iris scanners   Biometric devices that analyze the complex features of the iris, which is the colored part of the eye surrounding the pupil.

joint   The point of connection between two parts. A joint is the connection of movable parts to their parents.

keyframe animation   A type of animation where location and properties of key graphic elements are defined at critical junctures ("key" frames) in the animation. Software tools then create the intermediate frames by interpolating the changes in graphic elements between the keyframes. This process is also sometimes called "tweening" (short for in-between).

keyframing   A frame that clearly defines one or more constraining values. The animator determines changes in the scene at specific frames within the animation.

keystroke dynamics readers Biometric devices that measure the way a person types by analyzing keystroke patterns and speed. Keystroke dynamics are also called typing rhythms.

kilowatt (kW) The electric unit of power, which equals 1,000 watts or 1.341 horsepower.

kinetic energy A form of energy due to motion.

kinetic pumps Takes fluid near the central axis or rotation and, as the impeller rotates, the fluid is forced to the outside at high speed. This device can handle liquids, gases and fluids carrying small particles.

latent fingerprints Fingerprints left by sweat that is naturally produced by sweat glands present on the fingertips.

law of conservation of energy Law that states energy can neither be created nor destroyed, but can change forms.

lead acid batteries A type of battery often used to start cars.

line of sight An invisible projected plane from the viewer to an apparent object.

linear Long, straight, and narrow with parallel sides.

linear scale The lengths represented between each of the measurement reference points is equal.

load Used by heating and cooling engineers and technologists to describe the total amount of power needed to heat or cool a space. Also can be used to describe the capacity of a heating/cooling system.

lobe pumps Captures the liquid between a rotating lobe and the side wall of a chamber before it is pushed out the other side.

logarithmic scale A technique used in data display to provide maximum range while maintaining resolution at the low end of the scale.

macroscale The meteorological grid covering a large area. The scaled grid can range in size from a country, continent, or hemisphere.

magnetic energy A form of energy associated with the magnetic poles of the earth.

motion A change of position.

mainframe computers Used to describe a particular large computer system needing special space, power, network, and technical support. Modern mainframe computers are particularly useful for the management of large amounts of data, such as those used for accounting systems in banks or air traffic control systems.

match points Specific points of comparison.

mechanical energy A form of energy related to movement of objects as a result of an applied force.

megawatt (mW) Unit of electric power, used for measuring rate of producing or consuming electric energy. One megawatt = 1,000 kilowatts or 1 million watts. A megawatt is equal to 1,341 horsepower.

membrane electrode assembly (MEA) A component of fuel cells. The MEA separates the two flow field plates and is composed of several components, including the proton exchange membrane (PEM) and two electrodes.

metallic bonds Bonds where electrons essentially float free between the positively charged metal ions.

minutiae (*plural*) A variety of sweat pores, distance between ridges, bifurcations, junctions, and endpoints found in fingerprints. Minutia is singular.

misrepresentation To give a false or misleading representation of information or an idea, often with an intent to deceive or be unfair.

molecular assemblers   Machines that are molecular in size. These theoretical machines would be used to assemble additional assemblers as well as to build other small structures or objects.

molecular machines   Naturally occurring devices that perform specific functions for living things.

molecule   Two or more atoms held together by a chemical bond.

movable   Revolves relative to their parent part, usually to mimic the movement of the part in the actual product.

multimedia   A term used to refer to a system or process used to convey information in more than one media mode and often through more than one sense. An example would be a computer slide show that uses both text and graphics and has an audio track playing.

nano-   Prefix meaning one-billionth. Also used to refer to fields of nanoscience and nanotechnology in general.

nanomachine   A synthetic molecular machine produced through molecular manufacturing.

nanomanipulation   The ability to move atoms individually to achieve predetermined arrangements.

nanometer   One-billionth of a meter.

nanoparticles   A particle, microscopic in size, measured in nanometers.

nanoprobe   An example of a theoretical man-made molecular machine.

nanorobot   An example of a theoretical man-made molecular machine.

nanoscale   A scale that measures from hundreds to tens of nanometers, which is represented on a logarithmic scale as 10–9.

nanoscience   The study of nanoscale objects' fundamental physical, chemical, and biological properties.

nanoscientists   Scientists who study the fundamental physical, chemical, and biological properties of objects (e.g., atoms, molecules, and structures) at the nanoscale.

nanosensors   Sensors that work at the nanoscale.

nanoshells   Made of silicate or silver core nanoparticles surrounded by a gold coating.

nanostructure   A structure sized between molecular and microscopic structures. Nanostructures are the smallest solid devices possible to manufacture.

nanotechnologists   Professionals who apply the knowledge gained by nanoscientists to advance technology and develop applications using nanoscience.

nanotechnology   The creation of a material or machine through the manipulation of individual atoms and molecules through the application of nanoscience. Consists of the tools and techniques used to produce nanoscale objects and devices.

neutrons   Uncharged particle in the nucleus of an atom.

nickel cadmium batteries   A type of rechargeable battery used in a variety of consumer products.

nickel hydride batteries   A type of rechargeable battery often used in cellular phones.

non rigid transformations   Changes or distorts the shape of the object.

normal   When a face is at a 90° angle from the light source.

nuclear energy A form of energy that is released by a nuclear reaction, e.g., fission or decay.

object-oriented graphics A graphics system that manages graphic elements as coherent objects rather than pixels. For example, a user can change the color of a circle by changing the property of an object called "circle" rather than coloring individual pixels on the screen.

orbit The camera is wheeled in a circle around the object while keeping the camera pointed at the object.

orientation In graphics, the rotation of an object relative to other objects. The degree of rotation is typically calculated in degrees starting with a right-pointing horizontal vector.

orthographic Graphical depiction of a 3D object in 2D through the use of multiple views of the object.

oxidation–reduction or **redox** An electrochemical reaction where electrons are transferred from one chemical reactant to another. The chemical that loses an electron to the other substance is oxidized. The chemical that gains or accepts the electron is reduced.

pan To make a sweeping movement across the canvas or field of view.

parent Has children parts added later in the assembly and their location and/or movement is defined relative to this part.

particle systems A sophisticated type of procedural graphics that produces a set of similar objects.

parts Multiple individual geometric forms.

path-based animation Used for controlling translations and rotations of the model. With this approach, a curved path line is laid out in space, along which the model moves. The user can also specify how the model should rotate relative to the curving of the path line and, by indicating the number of frames for the path, how fast it should move along this line.

penstock A conduit carries water from the reservoir to the turbine in a hydroelectric plant.

peristaltic pump A positive displacement pump that uses rollers to move liquid by squeezing flexible tubing.

personal computer (PC) A type of computer that is small and simple enough that a single individual can manage it and use it in most any office or home location. The term is often comparable to desktop computers, but different from laptop or hand-held computers.

photoelectric effect The release of electrons from a material as a result of light striking its surface. This phenomenon eventually led the way for the development of photovoltaic cells.

photosynthesis The natural process used by green plants to convert sunlight into food using carbon dioxide and water.

photovoltaic cell A device that converts solar energy into electricity.

pictorial drawing also **pictorial views** A type of drawing where all three primary dimensions of the object are seen in a single drawing/view. Pictorial drawings are usually classified as either perspective or axonometric pictorials. Perspective pictorials are more realistic while axonometric pictorials are more commonly used in technical illustrations where distortion needs to be minimized.

piezoceramic tube scanner Holds the tip of the scanning tunneling microscope and moves the tip in motions small enough to be considered nanoscale movements.

piezoelectric materials   Ceramics that have been polarized by cooling down while in the presence of an electric field.

pigment   A substance that is applied to an object to give it color. Paints and printing inks both contain pigments. Pigments generate different colors by selectively absorbing certain wavelengths of light. Pigments are classified using many different color systems, such as Pantone™.

piston pump   Uses a piston in a cylinder to push fluids. The piston will have one side that encloses a chamber where the fluid comes in and leaves while the other side will usually be connected to an arm that controls the movement of the piston.

point light   Shines a uniform amount of light in all directions.

positive displacement pumps   Positive displacement pumps use forward pressure to move fluids. Such pumps include piston, plunger and screw pumps.

potential energy   Stored energy or energy due to position or structure.

primary colors   A set of colors from which all other colors can be generated. With computers and other light-based color systems, these colors are usually specific hues of red, green, and blue.

primitive instancing   An approach that allows users to specify the dimensions (parameters) of basic geometric solids.

prisms   2D polygons such as a rectangle, triangle, or hexagon that have been extruded into a third dimension. Predictably, rectangular, triangular, and hexagonal prisms are the results.

privacy   The ability to control the availability of information about yourself.

privacy invasion   When the right to be left alone is violated and unwanted intrusions occur.

procedural graphics   Any type of transformation of the model where the changes have been described as a series of procedures rather than defining exactly what the model should be doing in each frame.

profile   A plane in space on which to draw the 2D polygon. The plane on which the profile is drawn can be out in space or a flat surface on the model.

proton(s)   Positively charged particles.

proton exchange membrane (PEM)   The PEM is in the center of the membrane electrode assembly (MEA). On either side of the PEM is a thin layer of platinum catalyst.

pump   A device that moves liquid or gas from one space to another. A pump typically uses some type of motor to power a mechanism that moves the gas or liquid.

pumped storage (system)   A hydropower facility that has reservoir pumps that also serve as generators, installed in the dam.

pyramid   A shape in a geometrical form that is structured by joining triangular faces with a polygonal base and enclosing the end developing a point called the apex.

radiation   Electromagnetic energy emitted from a particular source. In the case of thermal radiation, it is a heat source. See also conduction and convection.

radiant energy   A form of energy from the movement of electromagnetic energy by transverse waves.

relative coordinate system   Cartesian frame that projects the most central frame. The axis is aligned with the direction of interest and positioned parallel to the most central vertical plane.

Renaissance   A transitional intellectual and artistic movement in Europe, beginning in fourteenth century Italy and ending in the seventeenth century. It is generally considered the transitional time between medieval and modern times, and it gave birth to modern science and technology.

rendering   The method of developing an image of a model by assigning surface properties, setting lights, and establishing a viewpoint.

respirocyte   A mechanical cell that mimics the actions of a natural red blood cell. Its size is roughly 1 micron in diameter.

retina   Light-sensitive membrane covering the back wall of the eyeball. Considered the primary organ of vision.

retinal scanners   Biometric devices that analyze the patterns formed by the blood vessels in the retina of the eye.

R-factor   A unit of measure used to describe a material's resistance to the passage of thermal energy through it. The higher the R-factor, the more resistance. A residential wall might have an R-factor of 20.

ribosomes   Molecular machines found within our cells that are responsible for building proteins.

ridges   Lines going across on finger pads.

rigid transformation   Rigid transformations do not change the shape of the object, only its location, orientation, or overall size.

rotating   Moving the object around the X, Y, and Z axes for an orientation change.

saturation   The ratio of the dominant hue in a color to all other hues. Can also be thought of as the dominant light wavelength relative to all other wavelengths. If a color is made up of only one wavelength, then it is fully saturated. If it is made up of an equal amount of all wavelengths, then it is completely desaturated—a shade of gray. As the saturation of a color increases, the color goes from a gray tone through washed out pastels, and on into rich, vibrant colors.

scaling   Will uniformly expand the object in the X, Y, and Z axes.

scanning probe microscopes (SPM)   Microscopes that map the surface of an object and use a computer to create a visual image based on data as measured or detected by the microscope.

scanning tunneling microscopes (STM)   Microscopes that work by measuring changes in the electrical current between the probe and the sample.

scene   All geometry created for a slide design in a virtual world is referred to as a scene.

scientific and technical visualization   A method or technique of creating graphic representations that communicate scientific or technical data and ideas. These visualizations can range from simple line graphs showing the power curve of a gasoline engine to sophisticated graphic representations of global warming.

screw pump   Screw pumps are revolving, positive displacement pumps that have at least one screw to transfer high or low thickness fluids.

secondary colors   A set of colors directly derived from mixing two primary colors. With a red/green/blue primary color system, the secondary colors are cyan/yellow/magenta.

self-assembly   The fundamental natural principle where disordered components (from atoms to galaxies) form structures or patterns without external intervention.

series   In data-driven visualizations, a series is all of the dependent variable data values associated with a particular independent variable level or value.

shearing   A part of an object is translated while the rest remains in place.

signature recognition systems   Biometric devices that analyze the features of a signature as well as the physical activity during the signature.

solar energy   A form of radiant energy that is derived from sunlight.

solar heating   A process that uses the sun's energy to derive and control heat.

solariums   Glass rooms attached to houses that collect and store the sun's energy to help indoor plants grow during cooler weather.

solid   An object that is composed of material from the innermost portion to the outermost layer.

sound energy   A form of energy derived from movement of energy through longitudinal waves, called sound waves.

specular reflection   Reflection of light at a designated angle from an objects mirror like surface.

sphere   A three-dimensional ball-shaped object where all points fall an equal distance from the center.

spot light   Shines a uniform amount of light in a single direction and focal angle.

spreadsheet   A specific type of software tool used extensively in the business, technical, and scientific communities to do basic mathematical calculations. Spreadsheets use a rectangular grid of cells, each of which can hold a simple numeric value, text, or the result of a mathematical calculation. Spreadsheets usually support the creation and display of data-driven visualizations, such as bar and line graphs.

static electricity   A form of electricity produced by friction.

static graphic   A type of graphic that does not change over time. See also dynamic graphic.

storyboards   A series of framed sketches or diagrams strung together used for planning projects. In animation, each framed sketch depicts a keyframe of the animation with text and arrow notations indicating what will change between one keyframe and the next.

stretching   Nonuniform scaling that increases or decreases the length of an object by scaling in a disproportional manner.

subassembly   When multiple parts, other than the base part, are added in as children to a parent part.

subtraction   A Boolean subtraction operation is when the new form is taken from the existing model. In a subtraction operation, the area of overlap between the forms is removed from the original model.

surface   The furthest layer of the outermost level of an object.

surface properties   Heavily depend on the points of each coordinate. Surface properties of a model include its color, the texture of its surface, and how that surface responds to light striking it. How the surface looks to the viewer will also depend on the number, types, and locations of the lights shining on the object and the viewpoint of the viewer.

sweeping   A 2D polygon is drawn and then moved along a linear or curved path to create a 3D solid.

syringe pump   A syringe is designed so that a plunger is driven down a single time by a shoe on a worm gear attached to gearing and a motor or directly to a stepping motor. To provide continuous fluid flow, the pump can be built as a double syringe unit where valves are used to allow one syringe to fill while the other is discharging.

system   A regularly interacting set of components whose coordinated activity achieves a larger, more sophisticated task than any one component would be capable of. The system can be mechanical (automotive power train system), organic (human digestive system), or cosmic (solar system), among others.

tapping mode   In an atomic force microscope, the tip works having intermittent contact across the surface.

telephoto   A focal length that provides a narrow field of visual coverage.

texture mapping   Formulating an image and inserting exterior qualities or appearances on its surface during rendering.

thermal energy   A form of energy from heat.

three-dimensional (3D, 3-D)   An object or a space that encompasses the three primary, orthogonal dimensions. These dimensions might have different labeling schemes, including X, Y, Z; height, width, depth; or vertical, horizontal, depth.

tilt   A vertical movement that shifts the field of view up or down.

timeline   Typically a two-dimensional graph that depicts key events at the time that they occurred. A historical timeline might indicate key historical events that happened on the date they occurred, while an animation timeline would indicate what graphic objects are doing on specific frames.

top-down approach   A nanotechnology technique that involves developing machining and etching techniques to create nanoscale structures.

torus   An object with a shape is similar to that of a donut. A torus is an elongated sphere with a hole central to the lengthened outer walls.

translated also **translations**   In computer graphics, the linear movement of an object in either two- or three-dimensional space along X, Y, and/or Z axes.

truck   The camera moves perpendicular to the direction of the line of sight or away from the subject.

turbine   Large blades that are turned by the force of water pushing against it; a turbine is connected to a generator.

tunneling   In a scanning tunneling microscope, the data points are gathered by exchanging electrons between atoms.

tweening   A process of interpreting data between two given values to structure an animation sequence. Software determined transformation scene from one keyframe to the next that is automatically generated for the in-between frames.

two-dimensional (2D, 2-D)   An object or a space that encompasses the two primary, orthogonal dimensions. These dimensions might have different labeling schemes, including X, Y; height, width; or vertical, horizontal. Relative to a viewer, these two dimensions tend to be normal (perpendicular) to the viewer's line of sight.

typing rhythms   The way a person types. Also called keystroke dynamics.

union   A Boolean union operation is when the new form is added from the existing model. With a union operation, any overlap between the two objects is only counted once.

valleys   Spaces between ridges on finger pads.

value   How light or dark a color is. In terms of rendering, value is determined by the amount of light striking that part of the model surface.

valves   When the piston is moving in and out, the valve is used for control when the chamber fills and empties.

variable   A way of defining a set of values. These values many be quantitative (1, 2, 3) or qualitative (red, purple, blue). In algebra, a variable is typically represented by a symbol such as *x*. In data-driven visualizations, variables are either dependent or independent.

verification   Process of making a match between a biometric trait and another piece of information. Also known as authentication.

viewpoint   The position from which you develop a perspective of an object.

virtual cameras   Used to give the user visual control over what the audience will view in a sequence.

virtual world   A computer-generated electronic representation of actual occurrences or events.

visible prints   Fingerprints left by fingers that are dirty or oily.

voice recognition systems   Biometric devices that authenticate vocal patterns by analyzing distinct patterns in a person's voice to verify the identity of an individual.

wide angle   A view of an object with a focal length of approximately 25 mm.

wind energy   A form of energy available from the movement of wind.

zoom   To change the viewer focus of an object to or from a lengthened view of the frame to or from a narrowed view of the frame.

zoom lens   A lens that allows the user to vary the focal range by adjusting from lengthened to narrowed views.

: